module 6

IGCSE

Mathematics

University of Cambridge Local Examinations Syndicate

Reviewed by **John Pitts**, Principal Examiner and Moderator for HIGCSE Mathematics

Edited by **Carin Abramovitz**

PUBLISHED BY THE PRESS SYNDICATE OF THE UNIVERSITY OF CAMBRIDGE
The Pitt Building, Trumpington Street, Cambridge CB2 1RP, United Kingdom

CAMBRIDGE UNIVERSITY PRESS
The Edinburgh Building, Cambridge CB2 2RU, UK http://www.cup.cam.ac.uk
40 West 20th Street, New York, NY 10011-4211, USA http://www.cup.org
10 Stamford Road, Oakleigh, Melbourne 3166, Australia
Dock House, Victoria and Alfred Waterfront, Cape Town 8001, South Africa

First published 1998
Fourth printing 2002

Printed by Creda Communications, Cape Town

Typeface New Century Schoolbook 11.5/14 pt

A catalogue record for this book is available from the British Library

ISBN 0 521 62514 9 paperback

Acknowledgements
We would like to acknowledge the contribution made to these materials by the writers and editors of the Namibian College of Open Learning (NAMCOL).

Illustrations by André Plant.

Contents

Introduction

Welcome to Module 6 of IGCSE Mathematics! This is the **sixth module** in a course of six modules designed to help you prepare for the International General Certificate of Secondary Education (IGCSE) Mathematics examinations. Before starting this module, you should have completed Module 5. If you are studying through a distance-education college, you should also have completed the **end-of-module assignment** for Module 5. The diagram below shows how this module fits into the IGCSE Mathematics course as a whole.

Module 1	Module 2	Module 3	Module 4	Module 5	Module 6
Assignment 1	Assignment 2	Assignment 3	Assignment 4	Assignment 5	Assignment 6

Like the previous module, this module should help you develop your mathematical knowledge and skills in particular areas. If you need help while you are studying this module, contact a **tutor** at your college or school. If you need more information on writing the examination, planning your studies, or how to use the different features of the modules, refer back to the **Introduction** at the beginning of Module 1.

Some study tips for Maths

- As you work through the course, it is very important that you use a **pen or pencil and exercise book**, and *work through the examples yourself* in your exercise book as you go along. Maths is not just about reading, but also about doing and understanding!
- Do feel free to write in pencil in this book – fill in steps that are left out and make your own notes in the margin.
- *Don't expect to understand everything the first time you read it.* If you come across something difficult, it may help if you read on – but make sure you come back later and go over it again until you understand it.
- You will need a **calculator** for doing mathematical calculations and a **dictionary** may be useful for looking up unfamiliar words.

Remember

- In the examination you will be required to give decimal approximations correct to **three significant figures** (unless otherwise indicated), e.g. 14.2 or 1 420 000 or 0.00142.
- Angles should be given to **one decimal place**, e.g. 43.5°. Try to get into the habit of answering in this way when you do the exercises.

The **table** below may be useful for you to keep track of where you are in your studies. Tick each block as you complete the work. Try to fit in study time whenever you can – if you have half an hour free in the evening, spend that time studying. Every half hour counts! You can study a **section**, and then have a break before going on to the next section. If you find your concentration slipping, have a break and start again when your mind is fresh. Try to plan regular times in your week for study, and try to find a quiet place with a desk and a good light to work by. Good luck with this module!

IGCSE MATHEMATICS MODULE 6

Unit no.	Unit title	Unit studied	'Check your progress' completed	Revised for exam
1	Statistics			
2	Probability			
3	Geometrical Transformations and Vectors			
4	Matrices and Matrix Transformations			

Unit 1
Statistics

The notion of **statistics** was originally derived from the word 'state' since it has been the function of governments to keep records of population, births, deaths, etc. The usual definition of statistics is the collection, organisation, summarisation, presentation and analysis of data. The average person conceives of statistics as columns of figures or zigzag graphs in the daily newspapers, associated with crime rates, exports and imports and the like. The general function of statistics is to develop principles and methods that will help us make decisions in the event of uncertainty. In fact, many people define statistics today as a method of decision-making in the face of uncertainty.

Statistics are widely used in our daily life. Scientists use them in their research. Industry uses statistics to plan future action. Insurance companies use statistics to fix the premium they must charge. Indeed, statistics have become an everyday tool of all types of professional people.

This unit is divided into four sections:

Section	Title	Time
A	Statistical tables and diagrams	3 hours
B	Averages	2 hours
C	Grouped and continuous data	2 hours
D	Dispersion and cumulative frequency	2 hours

By the end of this unit, you should be able to:

- organise data and present it in a frequency table
- construct pictograms, bar charts and pie charts
- construct histograms with equal intervals
- read, interpret and draw inferences from tables and diagrams
- find the mean, median and mode for individual and discrete data
- construct histograms with unequal intervals
- estimate the mean for grouped and continuous data
- draw up a cumulative frequency table
- draw a cumulative frequency graph
- estimate the median, percentiles, quartiles and inter-quartile range of a distribution.

A Statistical tables and diagrams

Stages of statistical investigation

Using statistics to aid our understanding of problems can, broadly speaking, be separated into three stages:

- Identifying the problem to be investigated and investigation of the problem.

- Collection of the data in the form of a list, a table or a graph.
- Studying the displayed data, drawing conclusions from it, and making decisions for the future.

Statistical data

Statistical data in the raw form may be presented in a table. Table 1 shows the marks of 25 students in a mathematics test in the order they are gathered.

Table 1

5	5	7	1	6
7	7	5	6	2
5	7	8	7	9
7	10	8	6	3
4	1	8	4	9

Since the figures are arranged in the order they were gathered, they are hard to interpret. If you are very good at figures you will be able to discover, after a few minutes, the maximum and minimum marks in the table. You may also know that you can obtain an 'average' of some kind by summing up the marks and dividing the sum by 25, the number of marks. But this is likely to be the only information you can squeeze out of the table.

To get more information and to get it quickly, you need to organise the data in some systematic way. The simplest way of doing so is to form an **array**. This is an arrangement of items according to their sizes. It can be formed either in ascending order (from the lowest to the highest value) or descending order (from the highest to the lowest value). Table 2 is an array of the 25 marks from Table 1. It is arranged in ascending order.

Table 2

1	1	2	3	4
4	5	5	5	5
6	6	6	7	7
7	7	7	7	8
8	8	9	9	10

Such an array has certain advantages over the data in the raw form. First, a glance at the array tells us that the range of marks is from 1 to 10. Second, we see clearly a great concentration of values near 7. Finally, the array reveals roughly the distribution pattern of the marks.

The array, however, is still a very cumbersome form of data organisation, especially when a large sample is involved. Moreover, its usefulness is exhausted after a few types of information have been obtained from it. It is therefore desirable to compress the data into a more compact form.

Frequency table

Raw results like these are normally organised into a table showing the number of times (**frequency**) each value occurs. This table is called a **frequency table**. This is an arrangement of data into columns (or rows). In the first column (or row) the numerical data are entered. So, using the example on the previous page, the first column would consist of the marks a student could obtain, listed from 1 to 10. The column labelled frequency shows the number of times each mark appeared.

Sorting or tallying

Every time an item appears, represent it by a vertical line (**tally mark**). It is convenient to mark each fifth item by a diagonal line cutting the previous four lines. For example:

When the item appears for the fizrst time draw one line. I

When the item appears for the second time add a line. II

When the item appears for the third time add a line. III

When the item appears for the fourth time add a line. IIII

When the item appears for the fifth time cross out the first four lines. ~~IIII~~

When the item appears for the sixth time draw one line and start the process once again. ~~IIII~~ I

In this way you can count in fives and the chance of mistake is reduced.

This is a frequency table using the information from Table 1.

Marks	Tally	Frequency
1	II	2
2	I	1
3	I	1
4	II	2
5	IIII	4
6	III	3
7	~~IIII~~ I	6
8	III	3
9	I	1
10	I	1

Graphs

Using a diagram to present data can help you to see what information you have. In order to allow visual interpretation of a distribution, it is common practice to draw a graph.

There are many possible ways to represent data in graphic form. Some of the forms are bar charts, pictograms, pie charts, histograms and line graphs.

You already know that to represent values on a graph, we use the two axes, the x-axis and the y-axis. Nonetheless, it is a good idea to mention some basic principles involved in presenting data by graphs:

- The object of the graph is to present the data as clearly and simply as possible. The reader will be confused if too many lines criss-cross each other. So minimise the lines that intersect each other.
- The axes must be carefully graduated and useful scales must be used. Unnecessary exaggeration or compression of the scale must be avoided. By choice of scales the same numerical values can be made to appear very different to the eye.
- The axes must be labelled appropriately.
- A correct title must be given to the graph.

Bar charts

In **bar charts** or **bar graphs** the information is represented by a series of bars (rectangles) all of the same width. The width of the bars doesn't matter, provided they are all of the same width. Sometimes the bars are just thick lines.

The bars can be touching or separated by small gaps of equal width. The height or the length of each bar represents the magnitude or frequency of the figures. The bars may be drawn vertically or horizontally.

A bar chart may be used to show more than one set of facts. This is useful if you want to compare two sets of facts.

Example 1

The table shows the number of pupils in a school who take part in various sporting activities. Represent these on a bar chart.

Sporting activities	Number of pupils
Soccer	120
Volleyball	80
Netball	100
Athletics	30
Rugby	10

Solution

Draw the two axes.
Label the horizontal axis 'Sporting activities'.
Label the vertical axis 'Number of pupils'.
Graduate the horizontal axis so that 5 small squares represent one game.
Divide the vertical axis so that 10 small squares represent 20 pupils.
Draw vertical bars (rectangles) with heights corresponding to the number of pupils and width equal to 5 small squares.

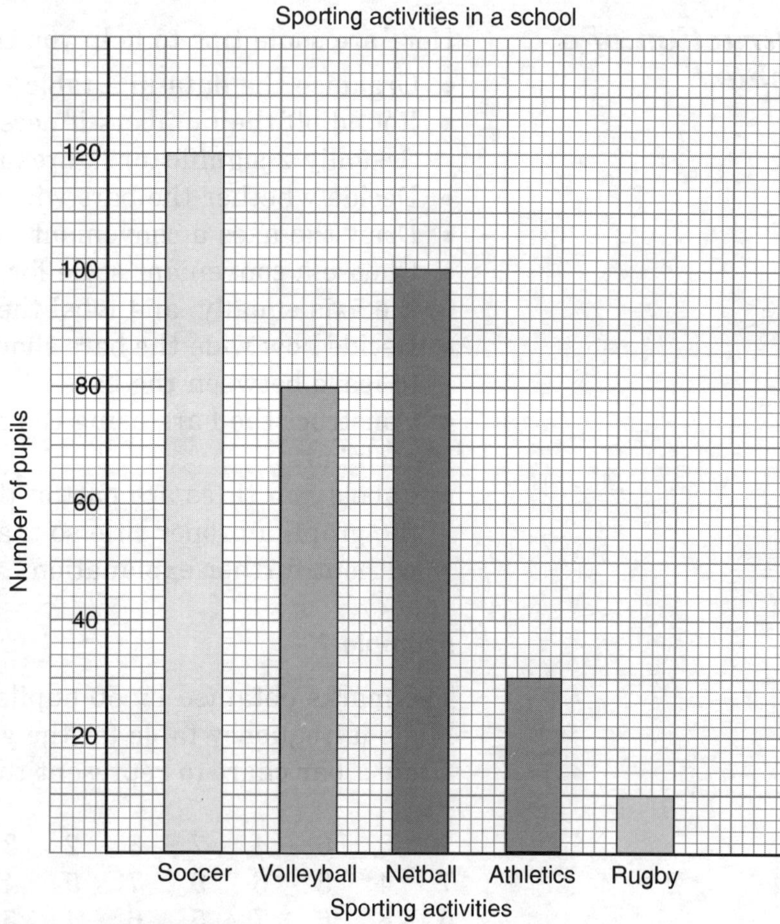

Example 2 and solution

The bar chart shows the population of a city.
Notice that from the graph we can only estimate the population. We cannot read them exactly.

The graph enables us to compare at a glance the populations in various years.

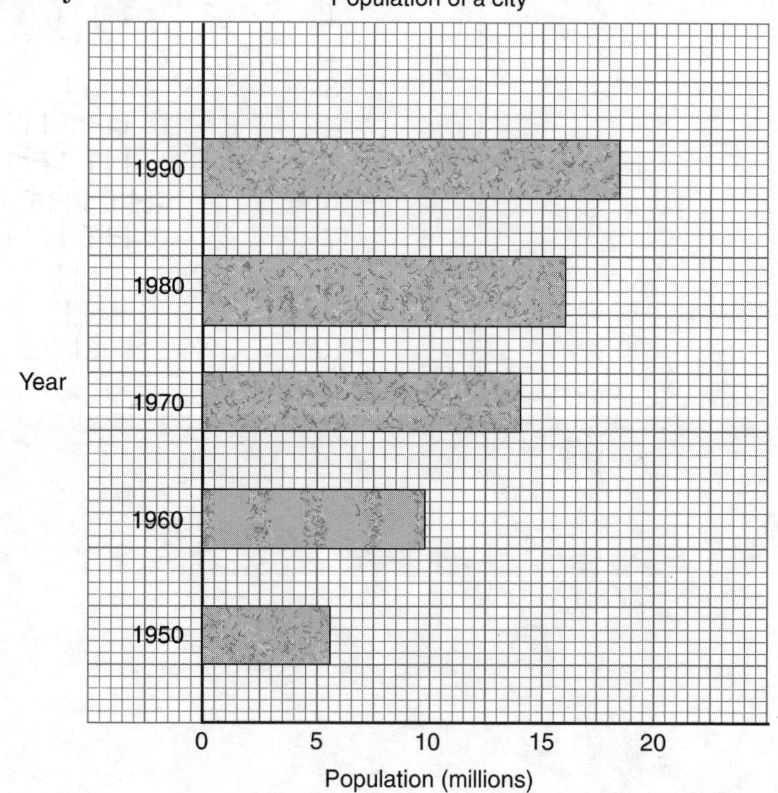

Construction of a bar chart

Here are some tips to help you construct a bar chart:

- Organise the data in a table.
- Round off the numbers if necessary.
  (Usually 2 significant figures is sufficient).
- Decide whether the bars will be vertical or horizontal.
- Draw the axes a convenient length.
- Choose a convenient scale for the vertical/horizontal axis, divide the axis equally, and label the division marks.
- Decide how wide the bars should be, and how much space (if any) to leave between the bars.
- Construct the bars using the appropriate height (or length) for each bar.
- Labels and titles are part of the graph. Label the axes and give the graph a proper title so that the graph can be understood without further explanation.

Example 1

The marks obtained by 50 pupils in a class test are given below.
Make a frequency table for the given marks.
Draw a bar chart to represent the given data.

10	3	6	4	7	8	2	3	4	1
7	4	5	6	9	7	5	4	6	7
5	8	6	7	5	6	4	5	6	8
5	6	7	5	4	7	5	6	1	6
6	5	6	9	1	5	4	6	7	7

Solution

Marks (out of 10)	Tally	Number (frequency)
1	III	3
2	I	1
3	II	2
4	~~IIII~~ II	7
5	~~IIII~~ ~~IIII~~	10
6	~~IIII~~ ~~IIII~~ II	12
7	~~IIII~~ IIII	9
8	III	3
9	II	2
10	I	1

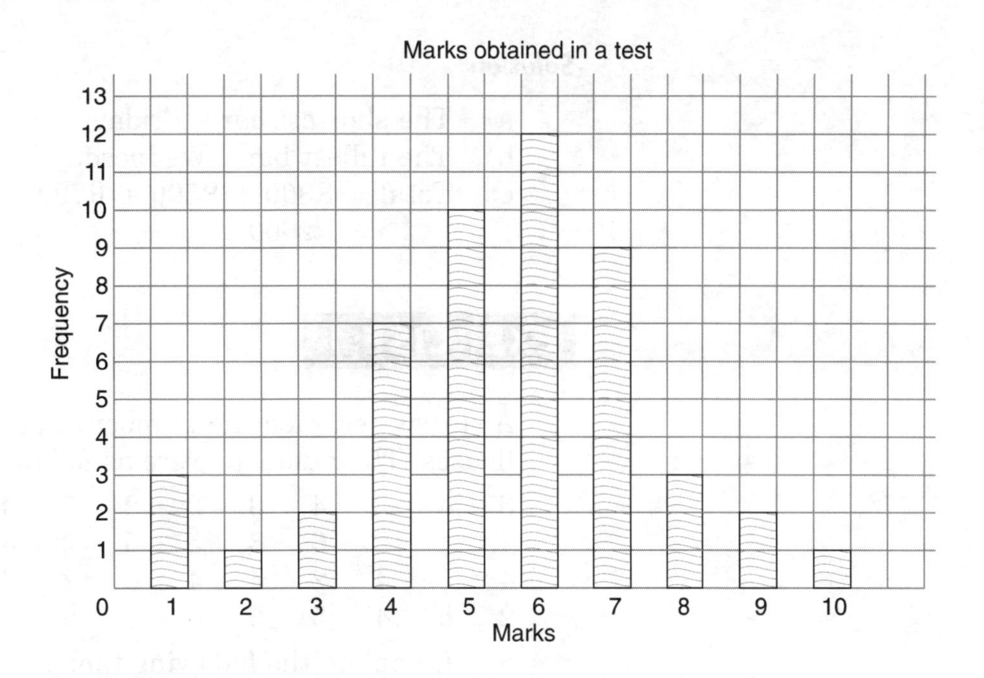

Example 2

The bar chart illustrates the daily takings of a small tuckshop.

a) On which day was the smallest amount of money collected?

b) On which day was the largest amount of money collected?

c) Find the total money collected for the week.

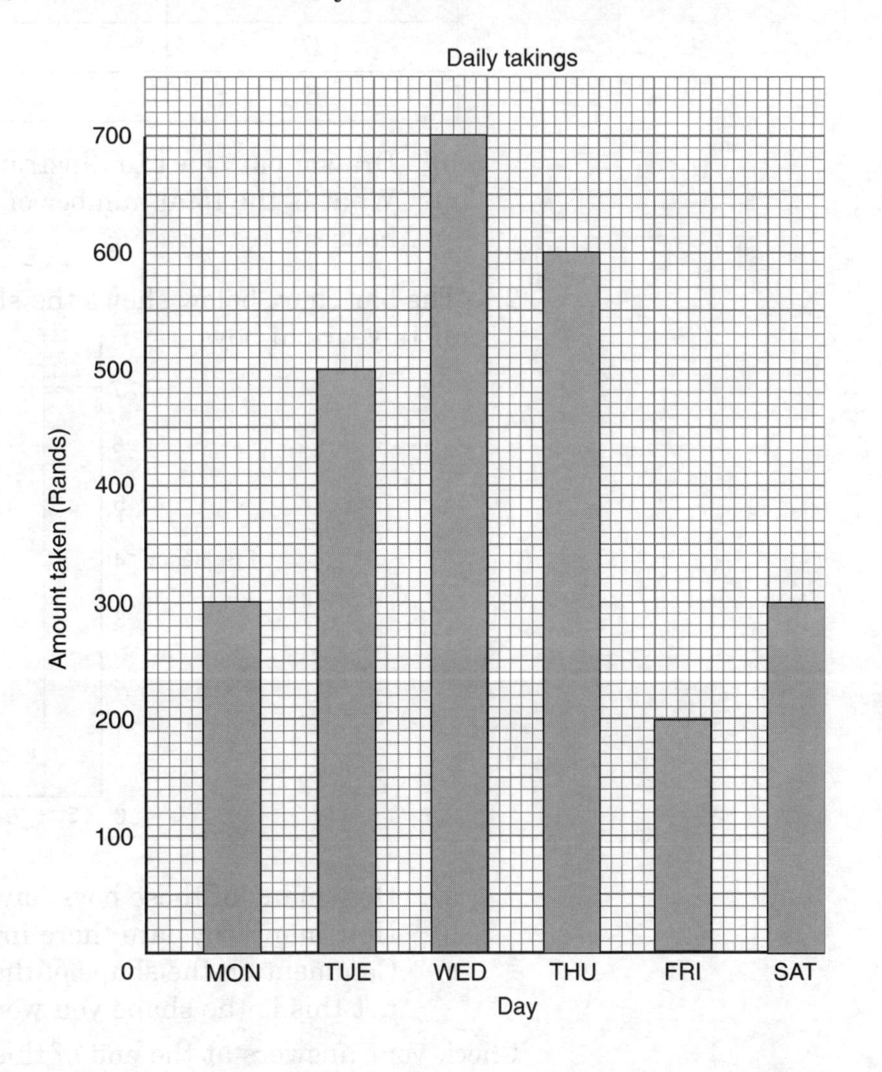

Solution

a) The shortest bar – Friday.
b) The tallest bar – Wednesday.
c) Total = R300 + R500 + R700 + R600 + R200 + 300
= R2600

EXERCISE 1

1. A survey recorded the number of people living in each of 40 houses. The numbers were as follows:

3	4	2	4	3	2	2	5	4	3
4	1	2	6	3	5	5	2	4	1
4	3	4	2	4	4	6	2	4	3
2	5	4	5	6	4	2	3	2	4

a) Complete the following table.

Number of people	Tally marks	Number of houses
1		
2		
3		
4		
5		
6		

b) Draw a bar chart to illustrate your results.
c) What is the total number of people living in these 40 houses?

2. The bar chart below shows the shoe sizes of a group of 16-year-old boys.

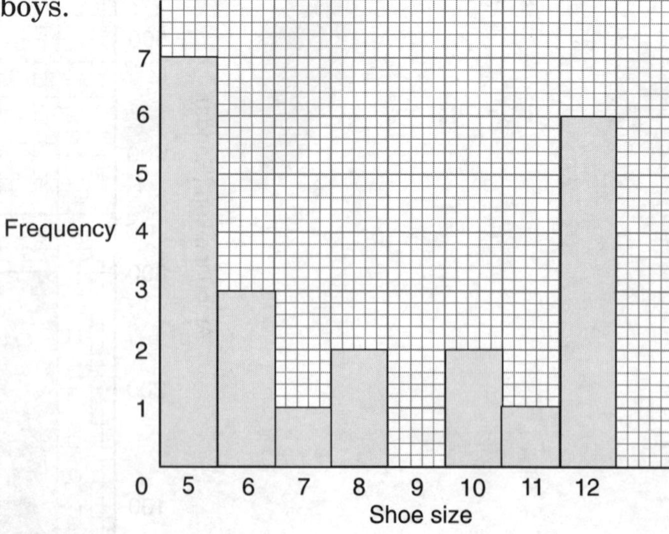

a) How many of these boys have a shoe size of 12?
b) How many boys are there in the group altogether?
c) Comment on the shape of the bar chart, saying whether or not this is the shape you would expect.

Check your answers at the end of this module.

Pie charts (circle graphs)

The **circle graphs** are commonly called **pie charts** because they resemble a pie which has been cut into sections.

In a pie chart a circle is divided into sectors corresponding in size to the data involved. The area of each sector is proportional to the angle at the centre of the circle. It is not easy to compare sectors with each other if they are nearly the same size. So the measurements are sometimes shown in the diagram.

Construction of a pie chart

- Add all the frequencies (items).
- Write each frequency as a fraction of the sum of all the frequencies.
- Change each fraction thus obtained into a number of degrees by multiplying by 360°.
  (The sum of the angles at the centre is 360°)
  For example, $\frac{3}{5} \rightarrow \frac{3}{5} \times 360^\circ = 216^\circ$
  The angles obtained are the angles of the sectors in the pie chart.
- Make a table showing the above data.
- Draw a circle of convenient radius.
- With a protractor, construct the successive angles at the centre corresponding to each sector.
- Label each sector.

Example 1

The table below shows how a student spends his time in a day.

Activity	School	Sleeping	Playing	Eating	Other
No. of hours	8	8	3	1	4

Show this on a pie chart.

Solution

- Sum of all the hours: $8 + 8 + 3 + 1 + 4 = 24$
- Change each number (hours) into a fraction of the whole (24 hours).
  School (8 hours): $\frac{8}{24}$; Sleeping (8 hours): $\frac{8}{24}$; Playing (3 hours): $\frac{3}{24}$;
  Eating (1 hour): $\frac{1}{24}$; Other (4 hours): $\frac{4}{24}$
- Change each fraction into a number of degrees.
  School: $\frac{8}{24} \times 360^\circ = 120^\circ$; Sleeping: 120°; Playing: $\frac{3}{24} \times 360^\circ = 45^\circ$;
  Eating: $\frac{1}{24} \times 360^\circ = 15^\circ$; Other: $\frac{4}{24} \times 360^\circ = 60^\circ$
- Set up a table.

Activity	School	Sleeping	Other	Playing	Eating
Angle	120°	120°	60°	45°	15°

- Draw a circle of radius, say, 2.5 cm as in Figure 1. Mark the centre of the circle.
- Draw 120° at the centre for the 'school' sector (Figure 2). Draw another 120° at the centre next to the 'school' sector for the 'sleeping' sector (Figure 3).

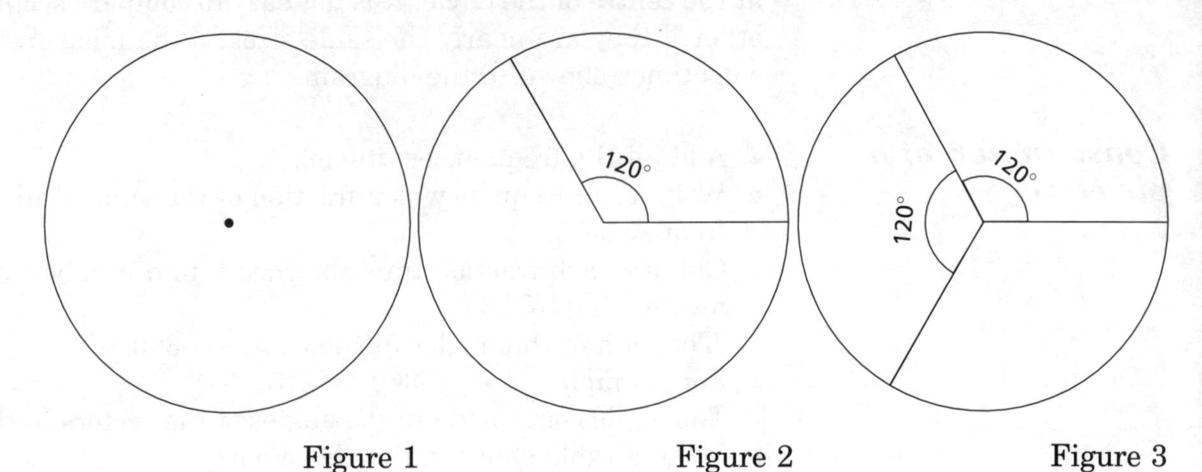

Figure 1 Figure 2 Figure 3

Complete the graph by drawing 60°, 45° and 15° as in Figure 4.
- Label the sectors as in Figure 5.

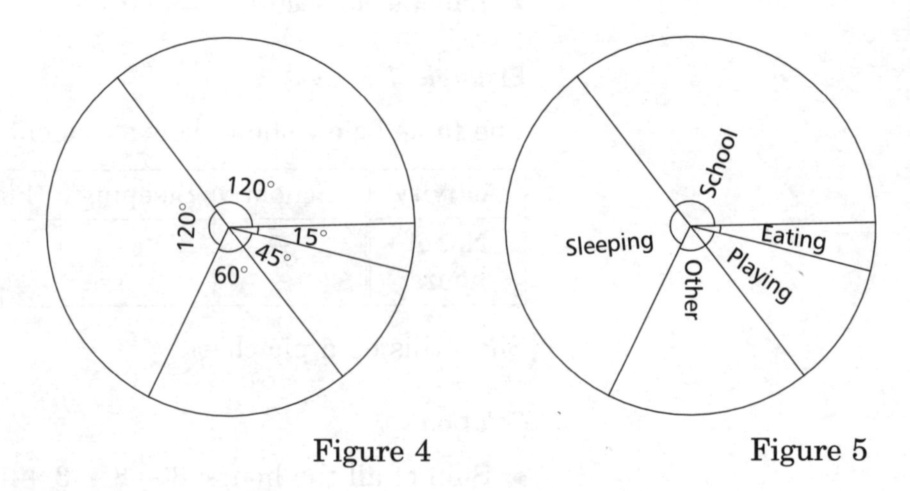

Figure 4 Figure 5

Example 2

A woman's expenditure per month is R1200 and is used as shown in the given pie chart.
The diagram is not drawn to scale.

a) If the angle at the centre in the transport sector is 90°, find the amount of money she spends on transport.
b) If she spends R700 on food, find the angle at the centre of this sector.
c) What fraction of her expenditure is on clothing?

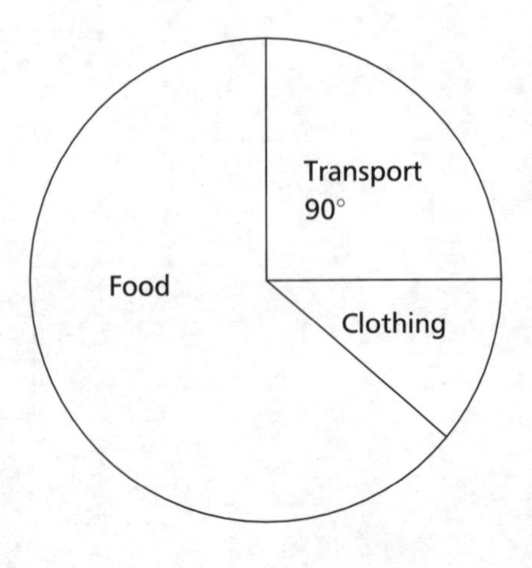

Solution

a) Sum of the angles at the centre $= 360^\circ$
Angle of the transport sector $= 90^\circ$

360° represents R1200

1° represents $R\frac{1200}{360}$

90° represents $R\frac{1200}{360} \times 90$

$= R300$

The amount she spends on transport = R300

b) R1200 is represented by 360°

R1 is represented by $\frac{360^0}{1200}$

R700 is represented by $\frac{360^o}{1200} \times 700 = 210^\circ$

Angle at the centre of the food sector $= 210^\circ$

c) Expenditure on clothing $= R1200 - R700 - R300$
$= R200$

Fraction of expenditure on clothing $= R\frac{200}{1200}$

$= \frac{1}{6}$

Pie charts are more difficult to draw and interpret than bar charts. Here is some practice for you.

EXERCISE 2

1. The given table shows the number of students in a university and their faculty studies. Draw a pie chart to illustrate this information.

Faculty	Students
Science	495
Arts	375
Medicine	108
Engineering	54
Law	48

2. The table shows how an income of R400 was spent. Show this data on a bar chart and a pie chart.

	Food	Rent	Clothing	Transport	Savings
Amount	R120	R80	R40	R110	R50

3. The diagram is a pie chart showing the expenses of a small manufacturing firm. The total expenses were R720 000. The angles at the centre of each sector are:
   Wages 150°; Raw materials 120°; Fuel 40°; Extras 50°.

   Work out how much was spent under each heading.

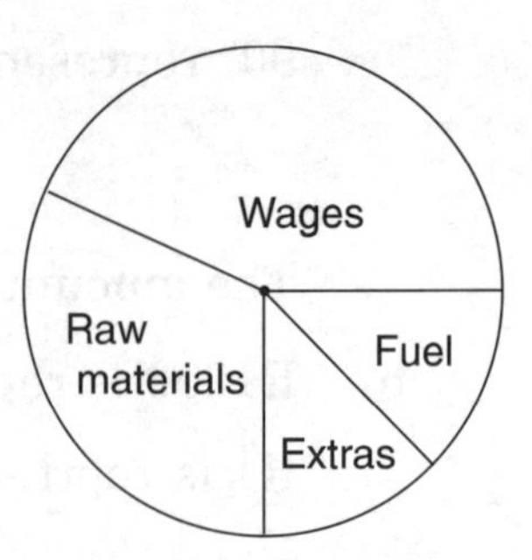

4. The pie chart shown below, which is not drawn to scale, shows the distribution of various types of land in a district.

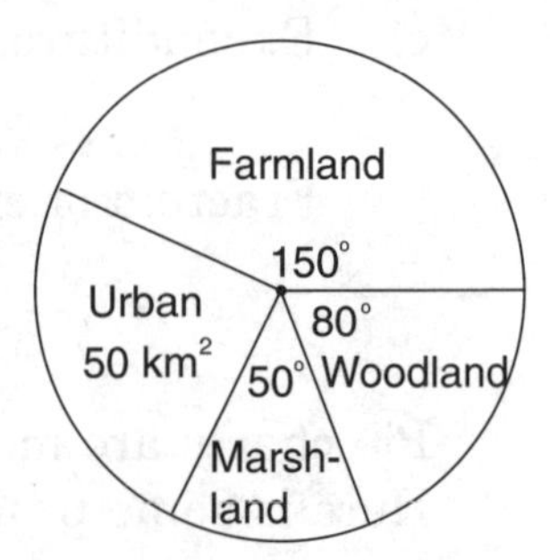

   Calculate:

   a) the area of woodland, as a fraction of the total area of the district
   b) the angle of the urban sector
   c) the total area of the district

Check your answers at the end of this module.

Pictograms

A **pictogram** is another kind of bar chart. Here the bars are replaced by a series of small pictures (called symbols or motifs).

Pictograms can provide a simple but striking display to illustrate certain data. They are mainly used in newspapers, magazines and reports of various sorts for those who are unskilled in dealing with figures or for those who have only limited interest in the topic depicted. They are suitable too, for comparisons rather than for measurements.

Construction of a pictogram

- Round off the numbers, if necessary. (Usually, 2 significant figures is sufficient.)
- Decide on a convenient unit to be represented by one figure.
- Select a simple figure to represent each unit. It should be one that can be easily drawn.
- Put a key at the bottom or side of the pictogram.
- Give a title for the pictogram.

Example 1 and solution

The diagram below is a pictogram representing the population of a country from 1930 to 1990.

Population of a country

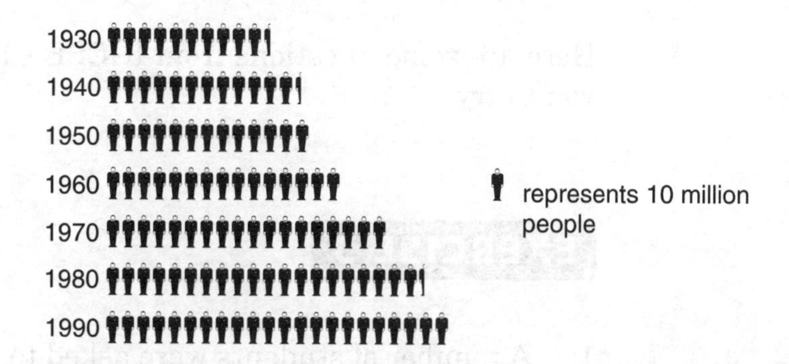

From the pictogram you can see that the population increases steadily during the years 1930 to 1990. In 1930 the population was about 105 million and in 1990 it was 220 million.

Example 2

The figure below shows a pictogram representing the mode of travel of students to a school.

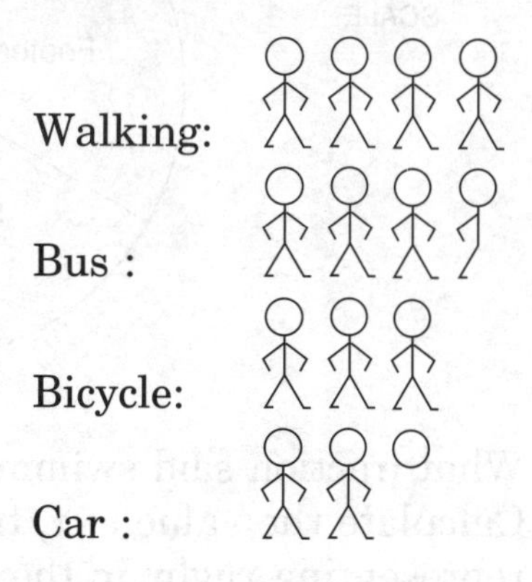

Key: represents 50 learners (head 10, arms and legs 10 each)

a) How many students go by car?
b) How many students go by bus?

Solution

a) Number of students travelling by car =

$= 50 + 50 + 10$
$= 110$

b) Number of students travelling by bus =

$= 50 + 50 + 50 +$ (head $+$ 1 arm $+$ 1 leg)
$= 150 + (10 + 10 + 10)$
$= 180$

Here are some questions from IGCSE examination papers for you to try.

EXERCISE 3

1. A number of students were asked to name their favourite sport. The results are shown in this pie chart. $\frac{1}{4}$ of the students said tennis, $\frac{1}{8}$ said rugby, $\frac{1}{3}$ said football and the rest said swimming.

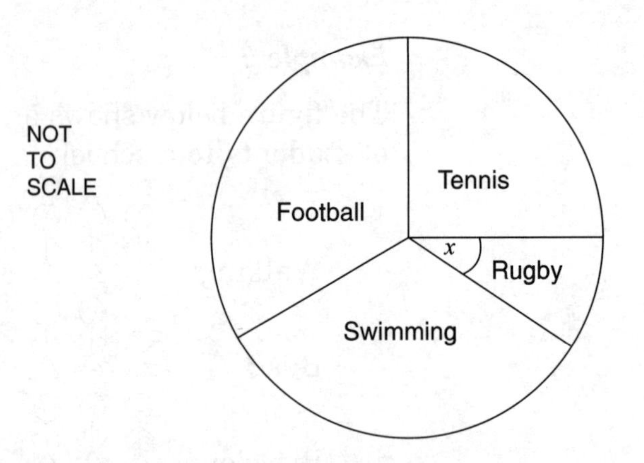

a) What fraction said swimming?
b) Calculate the value of x, the angle of the sector representing rugby in the pie chart.
c) If 32 students chose football, how many said tennis?

2. 48 football league matches were played on a Saturday. The pie chart shows the proportion of home wins, away wins and draws.

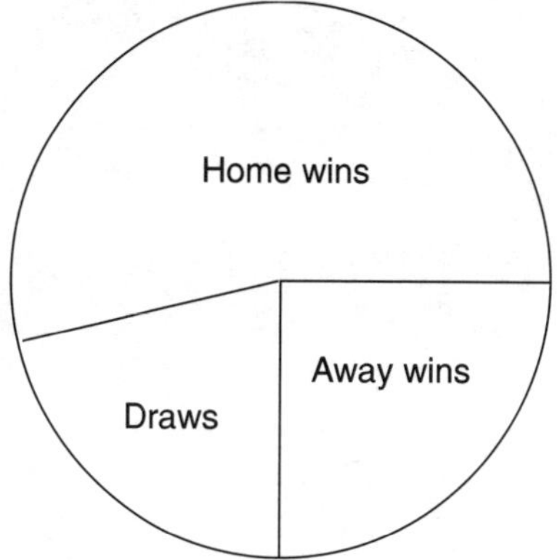

   a) Measure and write down the angle of the sector representing draws.
   b) Calculate the number of home wins.

3. The pie chart, which is accurately drawn, shows the nationalities of people staying in a holiday hotel.

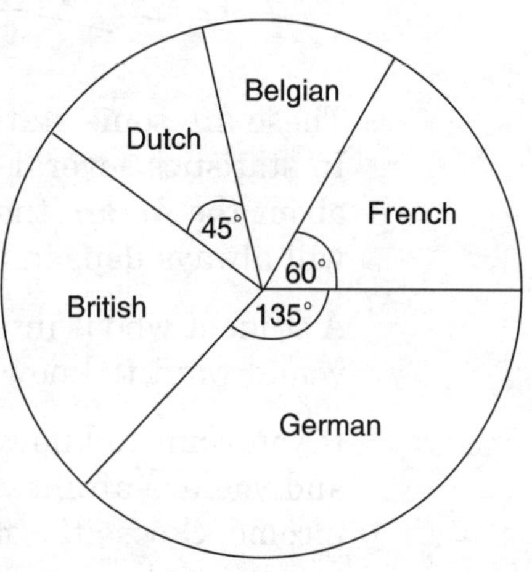

   a) Which of these five nationalities had the smallest number of people in the hotel?
   b) (i) What fraction of the people in the hotel were French? Give the answer in its lowest terms.
      (ii) Write the answer to b) (i) as a percentage correct to the nearest whole number.
   c) Write the ratio $\frac{\text{number of Germans}}{\text{total number of people}}$ as a decimal.
   d) If there were 288 people in the hotel altogether, how many of them were Dutch?

4. The pictogram shows the number of loaves of bread sold in a shop during each of three weeks. 440 loaves were sold in Week 1.

Week 1

Week 2

Week 3

Week 4

a) Estimate the number of loaves sold in Week 2 and in Week 3.
b) 370 loaves were sold in Week 4.
Complete the pictogram for Week 4.

Now compare your answers with ours which are at the end of this module. Are you confident that you can deal with this type of examination question?

B Averages

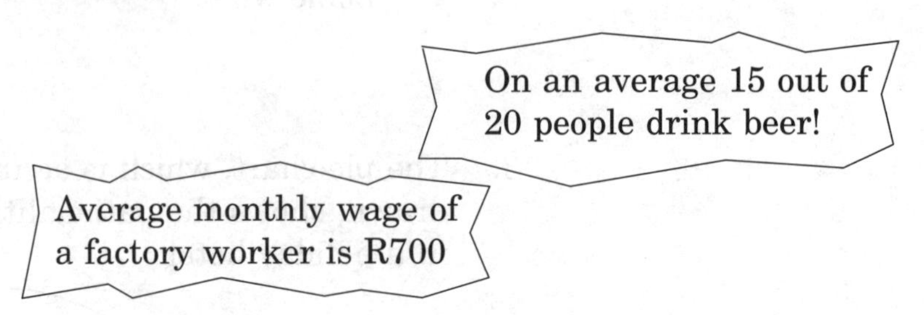

These are some statements you come across in everyday life. In statistics several kinds of average are used. Here we will learn about the **mean**, the **median** and the **mode**. The average one uses will always depend upon the purpose.

A student who is interested in his average performance in class tests would want to know the **mean**.

If you want to know the average income of people in a small town and you are afraid your scores will be influenced by a millionaire's income, choose the **median**.

A man who sells dresses will be interested in the size of dresses that are sold the most. Thus he will be interested in the **mode**.

The mean

This is the measure of average most often used. From the day you are born to the day you die, you are confronted with this measure. Your weight, height, test marks, etc. are associated with this single measure, which compares you with others. In fact, on many occasions you use it to find out information that deeply concerns you.

The mean is the number obtained by dividing the sum of all the items by the number of items.

$$\text{Mean} = \frac{\text{the sum of the items}}{\text{the number of items}}$$

Example 1

Find the mean of 8, 9, 7, 6, 8 and 10.

Solution

$$\begin{aligned}\text{The sum of the items} &= 8 + 9 + 7 + 6 + 8 + 10\\ &= 48\\ \text{The number of items} &= 6\\ \text{The mean} &= \frac{48}{6} = 8\end{aligned}$$

A formula for the mean

$$\bar{x} = \frac{X_1 + X_2 + X_3 \ldots X_N}{N}$$

where, the symbol $\bar{x}$ (which is read 'x bar') represents the mean and $X_1, X_2, X_3, \ldots X_N$ are the items. N represents the number of items.

The mean of a frequency distribution

The simplest type of frequency distribution is an arrangement of the items in order of size, with the frequency of each item indicated in a column labelled 'frequency'. You can find the mean using a frequency distribution table. Follow the examples to see how it's done.

Example 1

Calculate the mean for the following frequency distribution.

Test marks	Frequency
1	2
2	3
3	4
4	10
5	5
6	3
7	2
8	1
9	0
10	1

Solution

Draw 3 columns and label them, 'Test marks (X)', 'f' and 'fX'.
Enter the test marks in column 1.
Enter the frequencies in column 2.
Multiply the numbers in column 2 by the corresponding numbers in column 1 and enter the result in column fX.
Add the entries in column f.
Add the numbers in column fX.

Test marks (X)	f	fX
1	2	$2 \times 1 = 2$
2	3	$3 \times 2 = 6$
3	4	$4 \times 3 = 12$
4	10	$10 \times 4 = 40$
5	5	$5 \times 5 = 25$
6	3	$3 \times 6 = 18$
7	2	$2 \times 7 = 14$
8	1	$1 \times 8 = 8$
9	0	$0 \times 9 = 0$
10	1	$1 \times 10 = 10$
Total	31	135

$$\text{The mean} = \frac{135}{31}$$
$$= 4.35$$

what you have actually done here is found the total of all the marks and divided this by the number of marks

Example 2

Tickets for a circus have been sold at the following prices:
180 at R6.50, 215 at R8, 124 at R10.

a) What is the total amount of money received for the tickets?
b) What was the mean (average) price of the tickets sold?

Solution

Price of ticket (X)	Number of tickets (f)	fX
R6.50	180	R1170
R8	215	R1720
R10	124	R1240
Total	519	R4130

a) Total amount received = R4130
b) Mean price of the tickets sold $= \dfrac{\text{R}4130}{519}$
$= \text{R}7.95761$
$= \text{R}7.96$ to 3 significant figures

The median

The **median** of a number of items is the central or middle item when all the items are arranged in ascending (increasing) or descending (decreasing) order of size. In other words, half the items lie below the median and half the items lie above it, and the median therefore divides the distribution into two halves.

> For an odd number of items the median is the value of the item that is in the middle, after the items have been arranged in increasing or decreasing order.
>
> For an even number of items the median is the mean value of the two middle items, after the items have been arranged in increasing or decreasing order.

Example 1

Find the median of the following scores:

20 70 50 30 35 45 75 15 90

Solution

Arrange the scores in either ascending order or descending order.

Ascending order:

15 20 30 35 45 50 70 75 90

Descending order:

90 75 70 50 45 35 30 20 15

I always find it easier to arrange numbers in ascending order.
There are 9 scores.
The 5th score is the middle one.
The 5th score is 45.
The median is 45.

Example 2

Find the median of the following:

a) 10, 3, 4, 7, 8, 9, 1
b) 6, 5, 3, 8, 4, 2

Solution

a) Arranging in increasing order:
1, 3, 4, 7, 8, 9, 10

The number of items $= 7$

$n = 7$

$$\text{The median position} = \frac{(n+1)}{2} = \frac{(7+1)}{2} = 4$$

check that you agree that this little formula will give you the middle number

The number at the 4th position is the median.
The number at the 4th position is 7.
The median is 7.

b) Arranging in ascending order:
2, 3, 4, 5, 6, 8

The number of items $= 6$

$n = 6$

The median position $= \frac{(6+1)}{2} = 3.5$

The median is the mean of the values of the 3rd and 4th items.
The 3rd and the 4th items are 4 and 5.

The median $= \frac{(4+5)}{2} = 4.5$

Example 3

The distribution of marks obtained by the pupils in a class is shown in the table below.

Mark obtained	0	1	2	3	4	5	6	7	8	9	10
Number of pupils	1	0	3	2	2	4	3	4	6	3	2

Find the median of this distribution.

Solution

The total number of pupils $= 1 + 0 + 3 + 2 + 2 + 4 + 3 + 4 + 6 + 3 + 2$
$= 30$

$$\frac{n+1}{2} = \frac{30+1}{2} = 15.5$$

The median is the mean of the 15th and 16th marks when the marks are put in ascending (or descending) order.

We could list all 30 marks thus: 0 2 2 2 3 3 4 4 . . . but this is tedious, and it would be very time-consuming if the class contained a very large number of pupils.

An easier method is to add up the frequencies (numbers of pupils) in the table, starting at the left-hand end, until you reach the 15th and 16th pupils.

$1 + 0 + 3 + 2 = 6$ so the 6th pupil obtained 3 marks.
$6 + 2 = 8$ so the 8th pupil obtained 4 marks.
$8 + 4 = 12$ so the 12th pupil obtained 5 marks.
$12 + 3 = 15$ so the 15th pupil obtained 6 marks.

The 16th pupil is in the next group (of 4) and so obtained 7 marks.

The median $= \frac{6+7}{2} = 6.5$

The mode

The **mode** of a distribution is that number in the distribution that occurs most often. In other words, the mode is the item which occurs with highest frequency. A distribution which has two modes is called **bimodal**.

If there are more than two terms which appear most frequently, we will refer to such a distribution as having no mode. Such a distribution is called **nonmodal**.

The main advantages of the mode are that it requires no calculations, only counting and that it can be determined for qualitative (for example, colours), as well as quantitative data.

Example 1

Find the mode for the following distribution:

70 80 50 95 80 73 90 85

Solution

Arranging the numbers in ascending order:

50 70 73 80 80 85 90 95

Since 80 appears most frequently (twice), it is the mode.

Example 2

Find the mode of the following data:

3 6 3 5 3 8 1 8 5 4 2 8 10

Solution

Arranging the data in ascending order:

1 2 3 3 3 4 5 5 6 8 8 8 10

3 and 8 appear 3 times each.
3 and 8 are the modes.
This is a bimodal distribution.

Example 3

Find the mode of the given distribution.

Marks	Pupils
0	2
1	1
2	1
3	2
4	6
5	10
6	7
7	6
8	3
9	1
10	1

Solution

The highest frequency is 10. The mark against this frequency is 5.

Although the mode is a simple and useful concept, its applications present many troublesome aspects.

First, in a distribution two or more values may repeat themselves an equal number of times, and in such a situation there is no logical way of determining which value should be chosen as the mode. Secondly, we may not find any value that appears more than once. Thirdly, the mode is a very unsuitable value, because it can change radically with the method of rounding the data.

Finding the mean, median and mode of a set of readings is straightforward, but you must remember which is which! Try these questions.

EXERCISE 4

1. Construct a frequency table for the following data and calculate the mean.

   3 4 5 1 2 8 9 6 5 3 2 1 6 4 7 8 1
   1 5 5 2 3 4 5 7 8 3 4 2 5 1 9 4 5
   6 7 8 9 2 1 5 4 3 4 5 6 1 4 4 8

2. Find the median value of:

   a) 8, 1, 6, 7, 5, 2, 3
   b) 100, 75, 85, 95, 43, 99, 70, 60
   c) 2, 3, 1, 5, 6, 4
   d) 31, 28, 25, 21, 22, 20

3. Find the mode of each of the following sets of numbers.

   a) 4, 5, 5, 1, 2, 9, 5, 6, 4, 5, 7, 5, 5
   b) 1, 8, 19, 12, 3, 4, 6, 9
   c) 2, 2, 3, 5, 8, 2, 5, 6, 6, 5

4. A man kept count of the number of letters he received each day over a period of 60 days. The results are shown in the table below.

Number of letters per day	0	1	2	3	4	5
Frequency	28	21	6	3	1	1

   For this distribution, find:

   a) the mode
   b) the median
   c) the mean

Check your answers at the end of this module.

C Grouped and continuous data

Discrete and continuous data

Data can be classified as discrete or continuous. Discrete values occur in distinct steps, normally as whole units, such as the number of children in a family or number of houses in a village. For example, there are either 500 or 501 students in a school. There cannot be 500.3 or 500.4 students. There is no possibility of fractional value in this case. However, the values of a discrete variable need not be whole numbers.

Height, mass and age are examples of continuous data, since these qualities do not increase in steps. Measurements of continuous data approximate the true value. Usually continuous data are measured to the nearest whole unit. A given value in continuous data is regarded as ranging from 0.5 units below the given value to 0.5 units above the given value. Therefore, a mass of 68 kg would be considered to represent an interval from 67.5 to 68.5 kg entered for convenience as 68 kg.

Grouping of continuous data

The table below shows the heights of 50 students arranged in the ascending order. Each height was rounded up to the next whole centimetre, so a height of 141 cm means that the student's height, h cm, is in the range $140 < h \leq 141$.

131	132	133	134	134	135	135	136	136	136
136	137	137	137	138	138	139	139	140	140
141	141	142	142	142	142	142	143	143	144
144	144	145	145	145	146	147	147	147	148
148	149	149	149	150	150	151	151	152	153

We can group them, using h cm for the actual height, as follows:
$130 < h \leq 135$, $135 < h \leq 140$, $140 < h \leq 145$, $145 < h \leq 150$, $150 < h \leq 155$

Each group has a width of 5 cm.

A height which is a little more than 140 cm belongs to the group $140 < h \leq 145$, whereas a height of 140 cm belongs to the group $135 < h \leq 140$.

Height, h cm	Frequency
$130 < h \leq 135$	7
$135 < h \leq 140$	13
$140 < h \leq 145$	15
$145 < h \leq 150$	11
$150 < h \leq 155$	4
Total	50

Example 1 and solution

The table below shows the lengths of 50 pieces of wire used in a physics laboratory. Lengths have been measured to the nearest centimetre.

Length	26–30	31–35	36–40	41–45	46–50
Frequency	4	10	12	18	6

The interval '26–30' means 25.5 cm $\leq$ length < 30.5 cm.

Example 2 and solution

The following table shows the lengths of 50 telephone calls.

Length of call in minutes	0–	3–	6–	9–	12–	15–	18–
Frequency	5	6	8	11	8	5	7

The interval '6 –' means 6 minutes $\leqslant$ time < 9 minutes, so any time including 6 minutes and up to (but not including) 9 minutes comes into this interval.

The class boundaries are:

0 3 6 9 12 15 18

The final class boundary is not given. For calculation purposes, it would be taken as 21.

Example 3 and solution

The table below shows the masses of 50 small packets brought to a post office counter in a day.

Mass in grams	–100	–250	–500	–800
Frequency	8	12	20	10

The interval '–250' means 100 g < mass $\leq$ 250 g. This includes any mass greater than 100 g and less than or equal to 250 g.

The class boundaries are:

0 100 250 500 800

Example 4 and solution

The table below shows the ages of the teachers in a secondary school.

Age in years	21–30	31–35	36–40	41–45	46–50	51–65
Frequency	3	6	12	15	6	7

Ages are in completed years, not to the nearest year.
The interval '21–30' means 21 $\leq$ age < 31.
Someone who is 30 years and 11 months old would fall into this category.

Finding the mean of grouped data

- Find the mid-point of each interval (X).
- Multiply the frequency of each interval by its mid-point (fX).
- Find the sum of all the products fX.
- Find the sum of all the frequencies.
- Divide the sum of the products fX by the sum of the frequencies.

Example 1

The table below shows the lengths of 50 pieces of wire used in a physics laboratory. Lengths have been measured to the nearest centimetre.

Length	26–30	31–35	36–40	41–45	46–50
Frequency	4	10	12	18	6

Find the mean.

Solution

Find the mid-points of each interval.

Length	26–30	31–35	36–40	41–45	46–50
Frequency(f)	4	10	12	18	6
Interval	25.5-30.5	30.5-35.5	35.5-40.5	40.5-45.5	45.5-50.5
Mid-point(X)	28	33	38	43	48

Multiply f by X.

Length	26–30	31–35	36–40	41–45	46–50
Frequency (f)	4	10	12	18	6
Mid-point (X)	28	33	38	43	48
fX	$28 \times 4 = 112$	$33 \times 10 = 330$	$38 \times 12 = 456$	$43 \times 18 = 774$	$48 \times 6 = 288$

Find the totals.

Sum of $fX = 1960$

Sum of $f = 50$

The mean $= (1960 \div 50)$ cm

$= 39.2$ cm

Example 2

The table below shows the ages of the teachers in a secondary school.

Age in years	21–30	31–35	36–40	41–45	46–50	51–65
Frequency	3	6	12	15	6	7

Calculate the mean age of the teachers.

Solution

It is important to remember that ages are given in *completed* years. This means, for example, that '21–30' is the interval 21 years $\leq$ age $<$ 31 years with a mid-interval value of 26 years.

Age (years)	21–30	31–35	36–40	41–45	46–50	51–65
Mid-point (X)	26	33.5	38.5	43.5	48.5	58.5
Frequency (f)	3	6	12	15	6	7
fX	78	201	462	652.5	291	409.5

Sum of fX = 2094
Sum of f = 49

$$\text{Mean} = \frac{2094}{49} = 42.73469$$

Mean age of the teachers = 42.7 years to 3 significant figures.

Histograms

Histograms are like vertical bar charts. Unlike bar charts, in histograms there are no gaps between rectangles. Normally bar charts are used to represent discrete measurements and histograms are used to represent continuous data.

In histograms the *area* of each rectangle gives the number of items in the class interval. If all the class intervals have the same width, then all the bars have the same width and the frequencies can then be represented by the heights of the bars.

Example 1

Draw a histogram to illustrate the following frequency table. The measurements were made to the nearest unit.

Class interval	10–24	25–39	40–54	55–69	70–84	85–100
Frequency	1	4	11	5	3	3

Solution

The class boundaries are: 9.5 24.5 39.5 54.5 69.5 84.5 99.5
The class widths are: 15 15 15 15 15 15

Area of rectangle = class width × height of rectangle
As the class width is 15 for each interval,
area of rectangle = 15 × height of rectangle
So, area α height of rectangle
If we make the height of each rectangle the same as the frequency,
area α frequency.

So, we can draw the frequencies along the vertical axis and the class intervals along the horizontal axis.

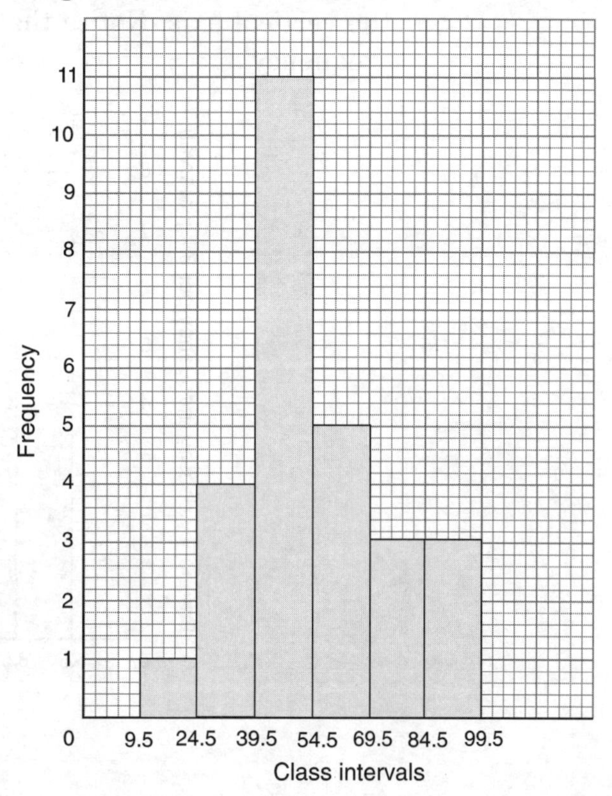

Example 2

The frequency distribution gives the masses of 35 objects, measured to the nearest kilogram.

Draw a histogram to illustrate the data.

Mass in kg	6–8	9–11	12–17	18–20	21–29
Frequency	4	6	10	3	12

Solution

The class boundaries are: 5.5 8.5 11.5 17.5 20.5 29.5
The class widths are: 3 3 6 3 9

The class widths are not equal. Therefore we cannot make the height of each rectangle equal to the frequency. So we choose a convenient width as standard and adjust the heights of the rectangles accordingly.

The first two class widths are 3 each and the heights of the first two rectangles can be taken as 4 units and 6 units respectively.
The third width is 6. So the height of this rectangle is half the frequency. The last interval is 9. This is 3×3. So the height of the rectangle is $\frac{1}{3}$ of the frequency.

In general, choose a width as 'standard' width.
If class width $= n \times$ standard width, then the height of rectangle $= \frac{1}{n} \times$ corresponding frequency.

Now draw the histogram with class boundaries along the horizontal axis and 'frequency density' (frequency per standard width) along the vertical axis. Notice that it is incorrect to label the vertical axis 'frequency'.

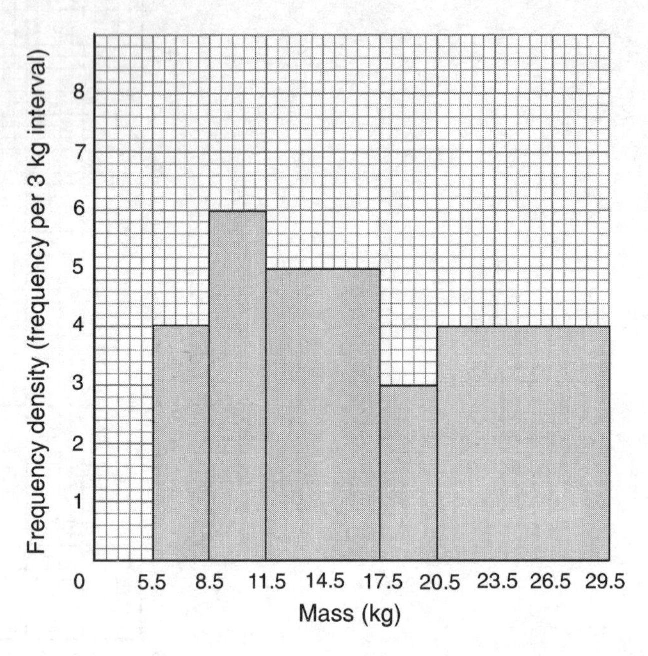

The mode of a grouped frequency distribution

The mode of an ungrouped frequency distribution is the item that has the highest frequency. We cannot find the mode for grouped data. We can only find out the **modal class** or modal classes. The modal class is the class interval which has the largest frequency. However, which class interval forms the modal class will depend on how the class intervals are chosen. An estimate of the mode can be obtained from the modal class.

Example 1

Find the modal class in the following distribution.

Marks	Frequency
6–15	2
16–25	7
26–35	9
36–45	3
46–55	4
56–65	2
66–75	1

Solution

The highest frequency is 9. The interval against this is 26–35.
So the modal class is 26–35.

To estimate the mode, first of all we must draw a histogram for the above data.

We can find an estimate of the mode by drawing lines as in the diagram. The estimated mode = 28 marks.

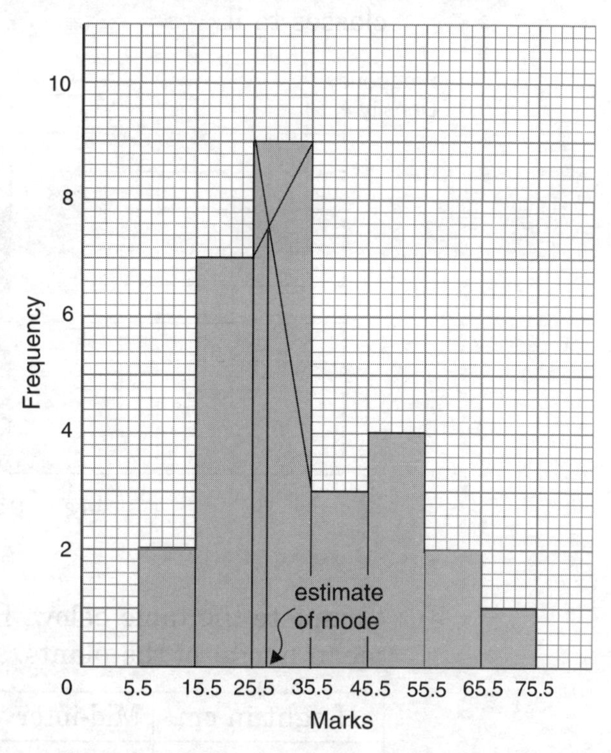

It is now time for you to answer some questions on grouped distributions.

EXERCISE 5

1. The frequency table shows the distribution of the masses of some objects. Draw a histogram to illustrate this distribution.

Mass (in grams)	$60 \le m < 63$	$63 \le m < 64$	$64 \le m < 65$
Frequency	9	12	15
Mass (in grams)	$65 \le m < 66$	$66 \le m < 68$	$68 \le m < 72$
Frequency	17	10	8

2. The heights of 25 plants were measured to the nearest centimetre. The results are summarised in the table.

Height in cm	Number of plants
5–14	4
15–19	8
20–24	7
25–39	6

a) The middle two classes are already shown in the histogram below. Complete the histogram by adding the other two classes to it.

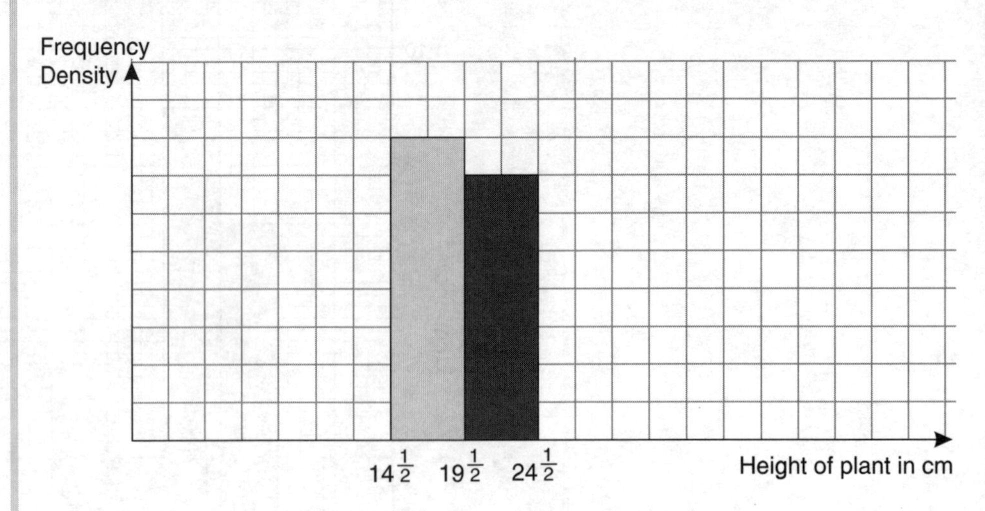

b) Complete the table below. Hence, or otherwise, calculate the mean height of the plants.

Height in cm	Mid-interval value (x)	Frequency (f)	fx
5–14		4	
15–19	17	8	
20–24	22	7	
25–39		6	

Check your answers at the end of this module.

D Dispersion and cumulative frequency

Dispersion

The average (mean, median or mode) gives us a general idea of the size of the data, but two sets of numbers can have the same mean but be very different in other ways.

The other main statistic we find is a measure of **dispersion** or **spread**. There are several ways of measuring dispersion.

Range

The **range** is the easiest measure of dispersion to calculate. It is defined as the difference between the highest value and the lowest value. The range is a crude measure of dispersion since it makes no use of the intermediate values and it can be distorted by one or two extreme values.

Inter-quartile range

This is a measure of the middle half of the data, so it is more representative.

A distribution is divided into four subgroups by three quartiles.

The first or lower quartile (Q_1) is the point below which 25% of the items lie and above which 75% of the items lie.

The second quartile (Q_2) is the point below which 50% of the items lie and above which 50% of the items lie. You will realise that the second quartile is the same as the median.

The third or upper quartile (Q_3) is the point below which 75% of the items lie and above which 25% of the items lie.

If there are n values, in ascending order, then the lower quartile Q_1 is the $\frac{(n+1)}{4}$th value, and the upper quartile Q_3 is the $\frac{3(n+1)}{4}$th value.

The **inter-quartile range** $= Q_3 - Q_1$.

Percentiles

The median and quartiles divide a distribution into four parts. For a large mass of data these may not give sufficient information. In such cases two other sets of measures are useful. They are **deciles** and **percentiles**.

Deciles divide the distribution into 10 equal parts.
Percentiles divide the data into 100 equal parts.

For example, the 10th percentile $P_{10} = \frac{10}{100}(n + 1)$th value.

The 90th percentile $P_{90} = \frac{90}{100}(n + 1)$th value.

The 10 to 90 percentile range $= P_{90} - P_{10}$.

Example 1

A student's marks in ten subjects in two sets of tests are given below.

Test 1:	20	22	28	19	20	24	23	20	24	20
Test 2:	13	15	36	11	18	30	23	8	32	34

Find, for each set of tests:

a) the range
b) the inter-quartile range
c) the median
d) the mean

Solution

a) Range in Test 1 = 28 – 19 = 9
Range in Test 2 = 36 – 8 = 28

b) In each case, the marks must be put in ascending order.

Test 1: 19 20 20 20 20 22 23 24 24 28

(the third value, 20, is the lower quartile; between the fifth and sixth values is the median; the eighth value, 24, is the upper quartile)

lower quartile ↓ median ↓ upper quartile ↓

Inter-quartile range for Test 1 = 24 – 20 = 4

Test 2: 8 11 13 15 18 23 30 32 34 36

13 ↓ lower quartile; 18, 23 ↓ median; 32 ↓ upper quartile

Inter-quartile range for Test 2 = 32 − 13 = 19

c) Median

Test 1: $\frac{20+22}{2} = 21$

Test 2: $\frac{18+23}{2} = 20.5$

d) Mean

Test 1: $220 \div 10 = 22$

Test 2: $220 \div 10 = 22$

Cumulative frequency table

In statistics we may have to answer questions such as "How many students scored more than 75%?" or "How many students scored less than 25%?".

To help us answer such questions it is convenient to add another column to the frequency table to show the total frequency up to and including the one corresponding to the value of the item we are interested in. That is, to show the frequency cumulatively.

The word 'cumulative' is related to the word 'accumulate' which means to 'pile up'. In other words, the **cumulative frequency** column shows the running total of the class frequencies. A cumulative frequency table gives the total up to each class boundary. This will be clear from the following example, which gives the marks obtained by 100 candidates in an examination.

The last number of the cumulative frequency column gives the total number of items or the sum of the frequencies.

The cumulative frequency column shows us that 2 candidates scored 9 or less, 6 candidates scored 19 or less, and so on. The successive entries are obtained by adding the next figure in the frequency column to the previous total figure. This is what is implied by the word 'cumulative'.

Score	Frequency	Cumulative frequency
0–9	2	2
10–19	4	6
20–29	8	14
30–39	10	24
40–49	12	36
50–59	25	61
60–69	22	83
70–79	8	91
80–89	6	97
90–99	3	100

Let us see how we obtained the cumulative frequency column in the example.

The entries in the cumulative frequency were obtained as follows.

1st entry: Enter the 1st class frequency (2).

2nd entry: Add the 1st class frequency (2) to the 2nd class frequency (4).
$2 + 4 = 6$

3rd entry: Add the 1st, 2nd and 3rd class frequencies.
$2 + 4 + 8 = 14$
Or, add the 2nd cumulative frequency entry to the 3rd class frequency.
$6 + 8 = 14$

4th entry: Add the 1st, 2nd, 3rd and 4th class frequencies.
Or, add the 3rd cumulative frequency entry to the 4th class frequency.
$14 + 10 = 24$

Continue this procedure to obtain the remaining entries.

Frequency	Cumulative frequency	Or	
2	2		2
4	$2 + 4 = 6$		$4 + 2 = 6$
8	$2 + 4 + 8 = 14$		$8 + 6 = 14$
10	$2 + 4 + 8 + 10 = 24$		$10 + 14 = 24$
12	$2 + 4 + 8 + 10 + 12 = 36$		$12 + 24 = 36$
25	$2 + 4 + 8 + 10 + 12 + 25 = 61$		$25 + 36 = 61$
22	$2 + 4 + 8 + 10 + 12 + 25 + 22 = 83$		$22 + 61 = 83$
8	$2 + 4 + 8 + 10 + 12 + 25 + 22 + 8 = 91$		$8 + 83 = 91$
6	$2 + 4 + 8 + 10 + 12 + 25 + 22 + 8 + 6 = 97$		$6 + 91 = 97$
3	$2 + 4 + 8 + 10 + 12 + 25 + 22 + 8 + 6 + 3 = 100$		$3 + 97 = 100$

If you complete the cumulative frequency column correctly, the last entry will be equal to the sum of the class entries.

Example 1

The heights of flowers were measured during an experiment. The results are summarised in the table.

Height (h cm)	$0 < h \leq 5$	$5 < h \leq 10$	$10 < h \leq 15$
Frequency	20	40	60
Height (h cm)	$15 < h \leq 25$	$25 < h \leq 50$	
Frequency	80	50	

a) Draw up a cumulative frequency table for this distribution.
b) Which class interval contains the median height?
c) Calculate an estimate of the median height of the flowers, correct to the nearest centimetre.

Solution

a) The entries in the cumulative frequency table will be
$20, 20 + 40, 20 + 40 + 60, 20 + 40 + 60 + 80,$
$20 + 40 + 60 + 80 + 50.$
The table is as follows:

Height (h cm)	$h \leq 5$	$h \leq 10$	$h \leq 15$	$h \leq 25$	$h \leq 50$
Cumulative frequency	20	60	120	200	250

b) Altogether, there are 250 flowers so the median is the mean of the heights of the 125th flower and the 126th flower.
There are 120 flowers with heights less than or equal to 15 cm, and 200 flowers with heights less than or equal to 25 cm.
Hence, the heights of the 125th and 126th flowers are each greater than 15 cm and less than or equal to 25 cm.
The median height of the 250 flowers is in the interval $15 < h \leq 25$.

c) There are 120 flowers in the interval $0 < h \leq 15$ and 80 flowers in the interval $15 < h \leq 25$.

The 125th and 126th flowers in the distribution are the 5th and 6th flowers (out of 80) in the interval $15 < h \leq 25$.

We estimate that their heights are $\frac{5}{80}$ths and $\frac{6}{80}$ths of the length of the interval from the lower end of the interval.

Estimate of the median of the distribution $= (15 + \frac{5.5}{80} \times 10)$ cm
$= (15 + 0.6875)$ cm
$= 15.7$ cm to 3 sig. figs.

Cumulative frequency curve

The curve obtained by plotting cumulative frequencies against the upper boundaries of the classes is called a **cumulative frequency curve**. (The curve is sometimes called an 'ogive' from a curve used in architecture.)

If the points are joined by straight lines, the graph is called a **cumulative frequency polygon**.

Example 1

The table below shows the examination marks of 300 students. Make a cumulative frequency table and draw the cumulative frequency graph. Find the median mark.

Mark	Frequency
1–10	3
11–20	7
21–30	13
31–40	29
41–50	44
51–60	65
61–70	70
71–80	49
81–90	14
91–100	6

Solution

Make the cumulative frequency table as explained earlier.
Draw a graph with cumulative frequency on the vertical axis against the upper boundary of each class on the horizontal axis.

I have drawn the graph on the next page.

Mark	Cumulative frequency
0	0
≤ 10	3
≤ 20	10
≤ 30	23
≤ 40	52
≤ 50	96
≤ 60	161
≤ 70	231
≤ 80	280
≤ 90	294
≤ 100	300

Total frequency is 300. The mark corresponding to the 150th candidate is the median mark.

Draw a line parallel to the mark-axis from the 150 mark on the cumulative frequency-axis so as to cut the graph.
Drop a perpendicular from the intersection point to the mark-axis.
This gives the median mark.
Median mark is 58.

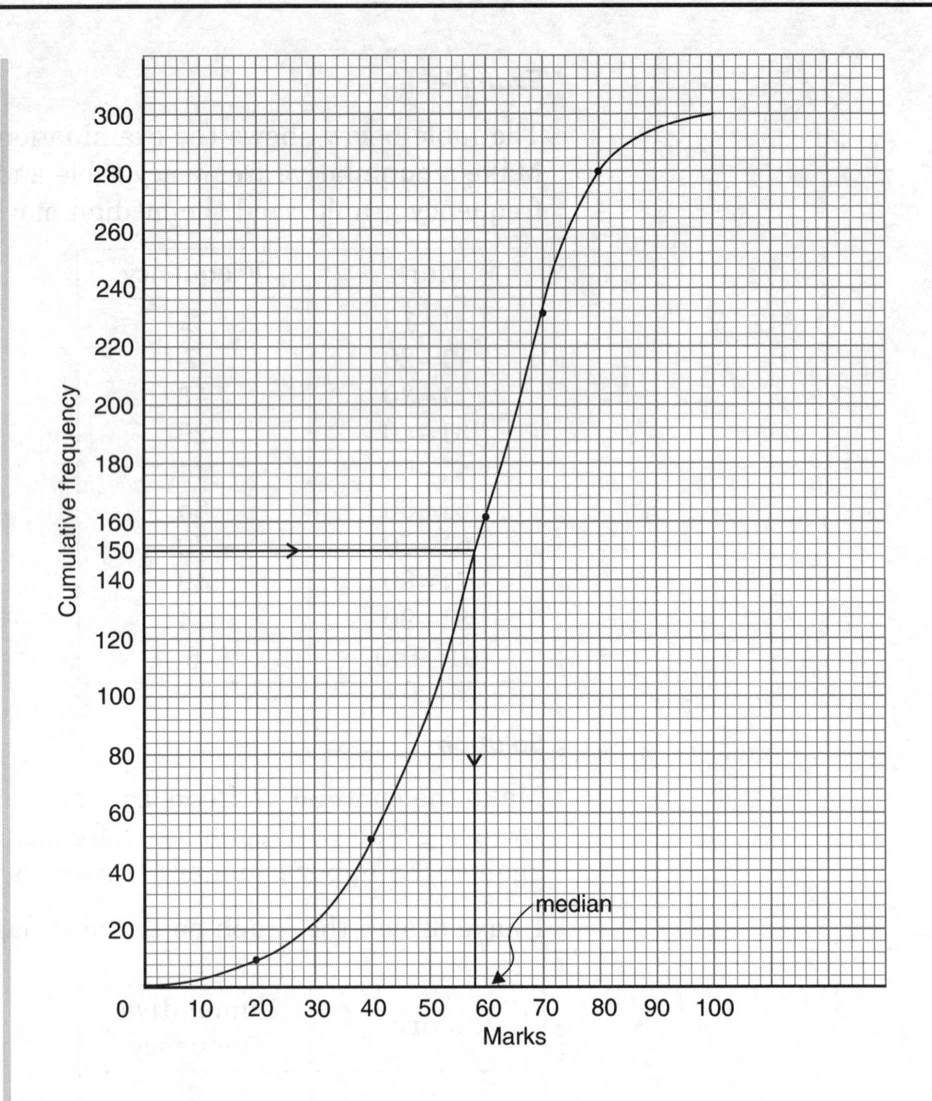

Example 2

The cumulative frequency curve shows the journey times to school of some students.

Use the curve to find:

a) the total number of students
b) the median journey time
c) the number of students who took less than 10 minutes to get to school
d) the number of students who had journey times greater than 30 minutes
e) the number of students who took between 40 minutes and one hour to get to school.

Solution

a) 50
b) 38 minutes
c) 4
d) $50 - 18 = 32$
e) $42 - 28 = 14$

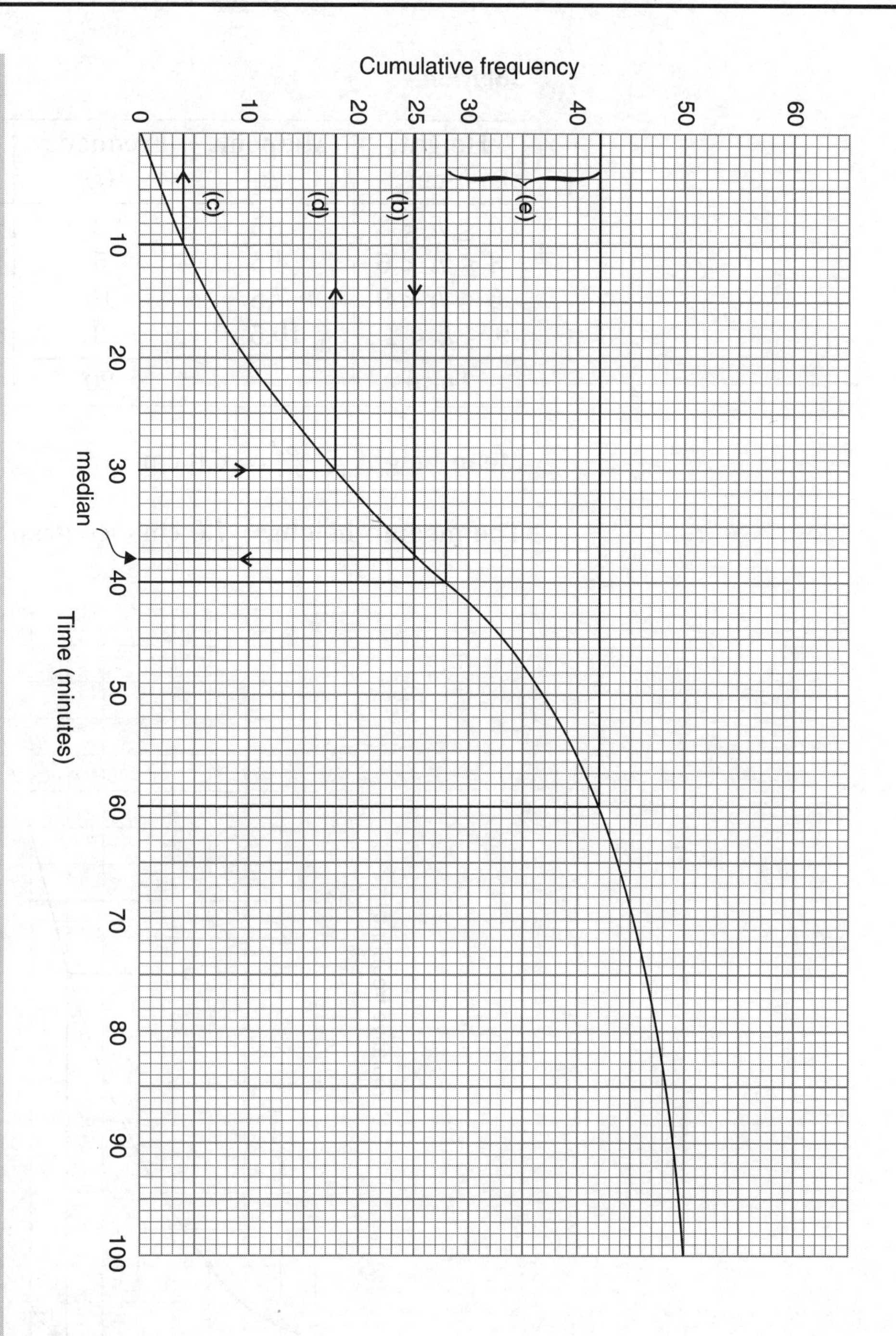

Example 3

Twenty bean seeds were planted for a biology experiment. The heights of the plants were measured after 3 weeks and recorded as below.

Find the mean height.

Draw a cumulative frequency curve and find the median height.

Height (h cm)	$0 \leq h < 3$	$3 \leq h < 6$	$6 \leq h < 9$	$9 \leq h < 12$
Frequency	2	5	10	3

Solution

Height, h cm	Midpoint (x)	Frequency (f)	fx	Cumulative frequency
$0 \le h < 3$	1.5	2	$2 \times 1.5 = 3$	2
$3 \le h < 6$	4.5	5	$5 \times 4.5 = 22.5$	7
$6 \le h < 9$	7.5	10	$10 \times 7.5 = 75$	17
$9 \le h < 12$	10.5	3	$3 \times 10.5 = 31.5$	20
Total		20	132	

Mean height $= \frac{132}{20} = 6.6$ cm

The median height = 7.0 cm (see graph)

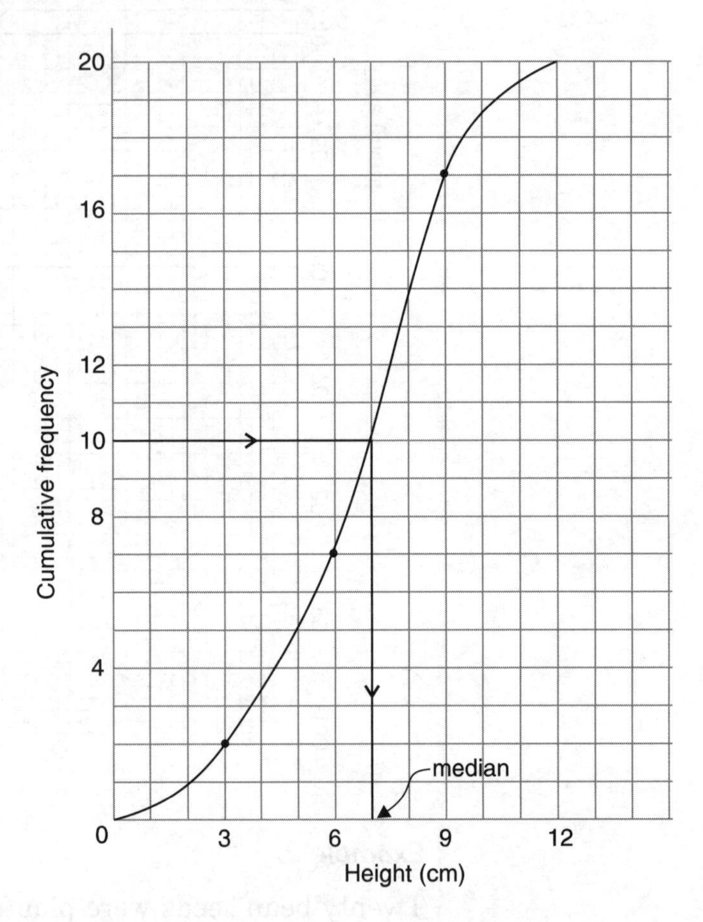

Questions on cumulative frequency usually take up a lot of time. Here are two questions similar to those you will find in IGCSE examination papers.

EXERCISE 6

1. The heights of 25 plants were measured to the nearest centimetre.
   The results are summarised in the table.

Height in cm	6–15	16–20	21–25	26–40
Number of plants	3	7	10	5

a) Draw up a cumulative frequency table for this distribution.
b) In which interval does the median plant height lie?
c) Estimate, to the nearest centimetre, the median plant height.

2. The table shows the amount of money, Rx, spent on books by a group of students.

Amount spent	$0 < x \leq 10$	$10 < x \leq 20$	$20 < x \leq 30$
No. of students	0	4	8
Amount spent	$30 < x \leq 40$	$40 < x \leq 50$	$50 < x \leq 60$
No. of students	12	11	5

a) Calculate an estimate of the mean amount of money per student spent on books.
b) Use the information in the table above to find the values of p, q and r in the following cumulative frequency table.

Amount spent	$x \leq 10$	$x \leq 20$	$x \leq 30$	$x \leq 40$	$x \leq 50$	$x \leq 60$
Cumulative frequency	0	4	p	q	r	40

c) Using a scale of 2 cm to represent 10 units on each axis, draw a cumulative frequency diagram.
d) Use your diagram:
(i) to estimate the median amount spent
(ii) to find the upper and lower quartiles, and the inter-quartile range

Check your answers at the end of this module.

Summary

This unit began by explaining how to draw up a frequency table to organise statistical data. You then learnt how to illustrate the information by means of:

- bar charts
- pie charts
- pictograms.

The three types of averages you need to remember how to calculate are:

- the mean $= \frac{\text{the sum of the items}}{\text{the number of items}}$
- the median which is the middle item
- the mode which is the item which occurs most frequently.

You should also know how to draw up a frequency table for continuous data and how to draw a histogram to illustrate the results.

I showed you how to find the mean and the modal class for grouped data.

The different techniques you need to know about for measuring the dispersion or spread of data are:

- the range = highest value − lowest value
- the inter-quartile range = upper quartile − lower quartile
- percentiles and deciles.

Finally, you learnt how to calculate cumulative frequencies and how to depict these on a graph.

In the next unit we'll be looking at the topic of probability and how this is measured. But first make sure you can do the questions in the 'Check your progress'.

Check your progress

1. Two hundred children were asked to choose their favourite pet. The results are represented in the pie chart (which is not to scale).
   a) Find the value of x.
   b) How many children chose rabbits?
   c) What percentage of children chose dogs?

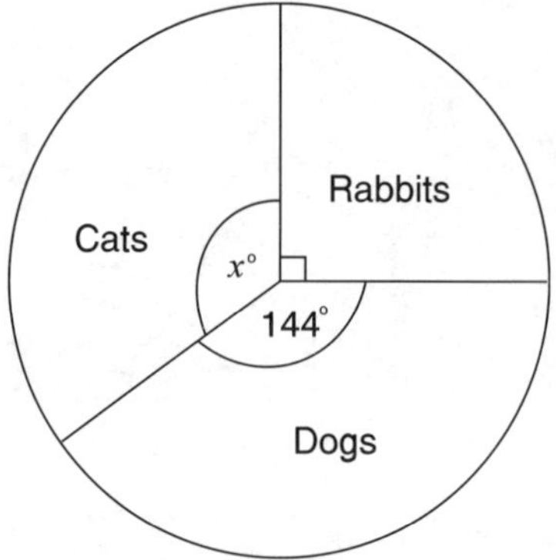

2. The bar chart shows the number of soccer teams scoring 0, 1, 2, 3, 4 or 5 goals on a particular day.
   a) How many teams scored 5 goals?
   b) How many teams were there altogether?
   c) How many goals were scored altogether?

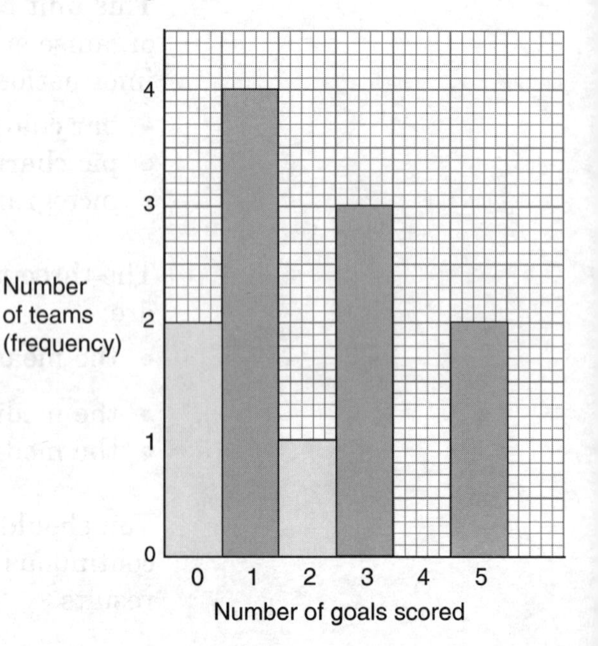

3. a) A bird-watcher recorded the number of eggs in each of 100 birds' nests.

Number of eggs per nest	1	2	3	4	5	6
Number of nests	5	18	28	34	9	6

Draw a bar chart to display these results.

b) Another bird-watcher looked into only 9 nests. He wrote down the number of eggs he saw in each as follows:

5, 1, 3, 6, 5, 5, 2, 3, 1

For this set of figures:

(i) find the mode
(ii) find the median
(iii) calculate the mean, giving your answer correct to 1 decimal place

4. Write down three positive integers with median 5 and mean 6.

5. After a morning's fishing, Arnold measured the mass, in grams, of each of the fish he had caught. The bar chart below represents his results.

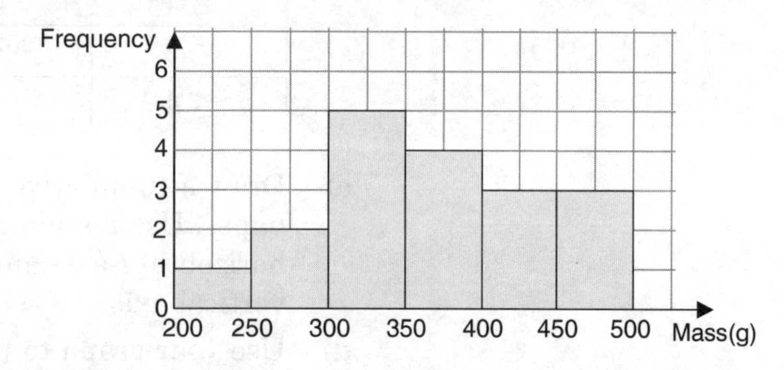

a) Use the bar chart to complete this table.

Mass (M) in grams	Number of fish	Classification
$M < 300$		Small
$300 \leq M < 400$		Medium
$M \geq 400$		Large

b) Represent the information in the table in a pie chart. Show clearly how you calculated the angles.

6. A survey of the number of children in 100 families gave the following distribution.

Number of children in family	0	1	2	3	4	5	6	7
Number of families	4	36	27	21	5	4	2	1

For this distribution, find:

a) the mode
b) the median
c) the mean

7. The table shows the length of time of 100 telephone calls.

Time (t minutes)	$0 < t \leq 1$	$1 < t \leq 2$	$2 < t \leq 4$
Number of calls	12	14	20
Time (t minutes)	$4 < t \leq 6$	$6 < t \leq 8$	$8 < t \leq 10$
Number of calls	14	12	18
Time (t minutes)	$10 < t \leq 15$		
Number of calls	10		

a) (i) Calculate an estimate of the mean time, in minutes, of a telephone call.
(ii) Write your answer in minutes and seconds, to the nearest second.

b) Make a cumulative frequency table for the 100 calls. Start like this:

Time (t minutes)	Cumulative frequency
0	0
≤ 1	12
≤ 2	26
≤ 4	

c) Draw a cumulative frequency diagram on a sheet of graph paper. Use a scale of 1 cm to represent 1 unit on the horizontal t-axis and 2 cm to represent 10 units on the vertical axis.

d) Use your graph to find, correct to the nearest 0.1 minute:
(i) the median time
(ii) the upper quartile
(iii) the inter-quartile range

I hope that you feel confident that you have a good grasp of the work in Unit 1. Check your answers at the end of this module.

Unit 2
Probability

When we know the probability of an event happening we can use its value to predict the likelihood of a future result. Government departments, business firms, industrialists, scientists, medical researchers and many other people use the figures from past events to predict what is likely to happen in the future, and thus they can plan ahead. That is why probability is linked to statistics.

This unit is divided into two sections:

Section	Title	Time
A	Probability of a single event	$1\frac{1}{2}$ hours
B	Combining events – part 1	$1\frac{1}{2}$ hours
C	Combining events – part 2	4 hours

By the end of this unit, you should be able to:

- find the probability of a single event
- draw and use tree diagrams
- calculate the probability of combined events.

A Probability of a single event

You may have heard statements like these:
"It is extremely likely that Black Tigers will win the National Cup."
"George Foreman has no chance of knocking out Mohammed Ali in the ring."
"The odds are against Russia beating America to Mars."

In all these statements there is a comparison of possible future events and the chance that one event is more likely to happen than another. We make these comparisons by using words and phrases such as 'almost certain', 'extremely unlikely', 'a good chance', 'probable', 'evens' and so on.

For many purposes these phrases are sufficient, but where a comparison has to be made which involves a payment of money, whether in betting or calculating the fire insurance to be paid on a house, it is necessary to be more exact. In mathematics we use the word **probability** for the happening (or not happening) of an event, but it is also known as 'chance' of an event.

Probability scale

Probability is measured on a scale from 0 to 1. A probability of 0 means that there is no chance of the event happening. A probability of 1 means that it is certain that the event will happen.

Look at the spinner given in the figure. Half of the spinner is painted black, one quarter of it is red and the other quarter is yellow. What is the probability that the pointer will stop on black?

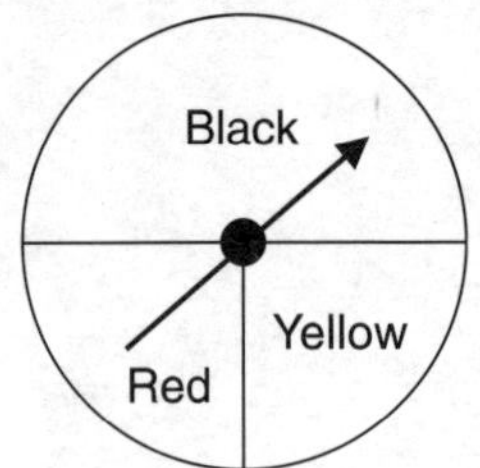

Since half of the spinner is black, the probability that the pointer will stop on black is $\frac{1}{2}$. Similarly, the probability that the pointer will stop on red is $\frac{1}{4}$.

A probability of $\frac{1}{2}$ is sometimes described as a 50-50 chance of the event happening.

The nearer the value of the probability is to 1, the more chance there is of the event happening.

Probabilities can be illustrated on a number line as given below.

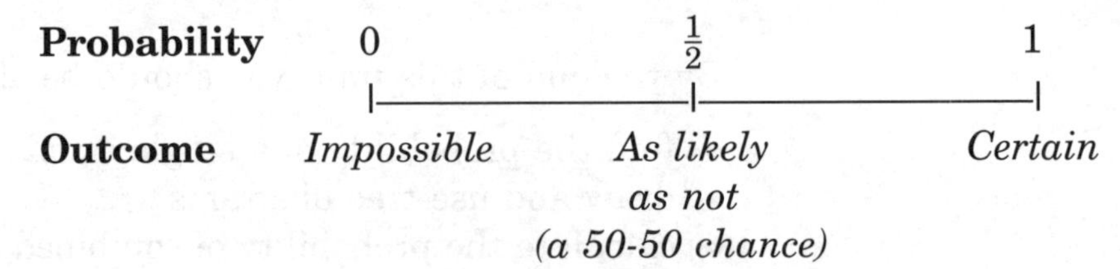

Experimental probability

This is a die. The plural of die is dice.

In an experiment a die was thrown 72 times. The number on its face showing up with the number of times it showed up was recorded as follows:

Number on the face showing up	Number of times
1	14
2	12
3	8
4	13
5	9
6	16
	Total = 72

In another experiment the same die was thrown 240 times and the results were as follows:

Number on the face showing up	Number of times
1	40
2	39
3	44
4	38
5	37
6	42
	Total = 240

In the first experiment, 6 was recorded twice as often as 3. Does this mean that the chances of throwing '6' with the die are twice as good as those of throwing '3'? No, because in the second experiment, 6 was scored 42 times and 3 was scored 44 times.

From the second experiment we notice that each of the scores 1, 2, 3, 4, 5, 6 came up approximately the same number of times. That is, 40 times each out of 240 times. In the second experiment, 1 has come up 40 times out of 240 throws. We say that the experimental probability of 1 showing up is $\frac{40}{240}$. In general, the more times the die is thrown, the nearer the experimental probability will be to the actual probability.

Try a similar experiment with a drawing pin.
Toss a drawing pin 50 times.
Record whether the pin lands with its 'pin up', or 'pin down'.

Repeat the experiment for 500 and 1000 tosses. What do you estimate to be the probability of the outcome 'pin up'? Of 'pin down'? What do you notice about the sum of these two probabilities?

> Experimental probability of an event happening
>
> $$= \frac{\text{the number of times the event happened}}{\text{the total number of trials}}.$$

Expected or theoretical probability

Earlier, we were able to determine the probability of an event happening by carrying out a series of experiments. It is likely that someone else carrying out the same experiments might obtain a different value for the probability although they are usually approximately the same. Sometimes it is possible to deduce the probability of an event happening by quite different means.

When an unbiased coin is tossed a large number of times, there is no reason why more 'heads' should turn up than 'tails'. We say 'head' or 'tail' is *equally likely* to turn up when an unbiased coin is tossed. Similarly, when a 'fair' die, a perfectly symmetrical cube, not loaded, is rolled, each of the six numbers on it are *equally likely* to turn up. When we see no reason why one outcome of an experiment should occur more frequently than another, we say that the outcomes are **equally likely**.

Consider a die which consists of a perfectly symmetrical cube. Because of the symmetry it is reasonable to say that the probability of '1' turning up is the same as that of '2' turning up, and so on. There are six ways in which a die can land and, since these are equally likely to happen, the probability that any particular number on a die will turn up is $\frac{1}{6}$.

This is the **expected** or **theoretical probability** of throwing a particular number with a die, and the experimental probability usually approximates to it.

We may define the probability as follows:

$$\text{The probability of a favourable outcome} = \frac{\text{the number of favourable outcomes}}{\text{the number of possible equally likely outcomes}}.$$

Example 1

Find the probability of a tossed coin (unbiased) showing heads.

Solution

There are two equally likely outcomes, heads and tails.
Therefore, the number of possible equally likely outcomes $= 2$.
The favourable outcome is heads.
That is, the number of favourable outcomes is 1.
Probability of heads $= \frac{1}{2}$.

Example 2

When an unbiased die is rolled, what is the probability that an even number turns up?

Solution

The possible equally likely outcomes are 1, 2, 3, 4, 5, 6.
Therefore, the number of possible equally likely outcomes $= 6$.
The even numbers are 2, 4, 6.
That is, the number of favourable outcomes is 3.
Hence, the probability that an even number turns up $= \frac{3}{6} = \frac{1}{2}$.

Example 3

If a card is drawn at random from a pack of playing cards, what is the probability that it is an ace?

Solution

There are 52 cards in a pack of cards.
Therefore, the number of possible equally likely outcomes $= 52$.
In a pack of cards there are 4 aces.
That is, the number of favourable outcomes $= 4$.
Therefore the probability of obtaining an ace $= \frac{4}{52} = \frac{1}{13}$.

Example 4

A bag contains 10 red and 6 blue beads, identical apart from colour.
A bead is drawn at random from the bag.
What is the probability that it is a red bead?

Solution

There are $10 + 6 = 16$ beads.
So we can draw any one of 16 beads.
Therefore, the number of possible equally likely outcomes = 16.
We can draw any one of the 10 red beads.
That is, the number of favourable outcomes = 10.

Therefore, probability of drawing a red bead $= \frac{10}{16} = \frac{5}{8}$.

Example 5

The foot sizes of the 11 players in a women's hockey team are shown in the table.

Foot size	38	39	40	41
Number of players	2	5	3	1

One of the players is chosen at random.
What is the probability that her foot size is 40?

Solution

Number of players in the team $= 2 + 5 + 3 + 1 = 11$.
Of these 11 players, 3 have foot size 40.
Probability that a player, chosen at random, has foot size 40 $= \frac{3}{11}$.

Example 6

a) Margaret chooses a card at random from the six cards shown in the diagram.

A B A C U S

What is the probability that the card is:

(i) an A
(ii) an A or a U
(iii) an E

b) From the six cards, card B is removed. Paul chooses a card at random from the five remaining cards.
What is the probability that the card is an A?

Solution

a) (i) The six cards are equally likely to be chosen and two of them are As.
Probability that the card chosen is an A $= \frac{2}{6} = \frac{1}{3}$.
(ii) Of the six cards, three are A or U.
Probability that the card chosen is an A or a U $= \frac{3}{6} = \frac{1}{2}$.
(iii) Of the six cards, none is an E.
Probability that the card chosen is an E $= \frac{0}{6} = 0$.

b) The five remaining cards are A, A, C, U, S.
These five cards are equally likely to be chosen and two of them are As.
Probability that the card chosen is an A $= \frac{2}{5}$.

Probability that an event does not happen

A bag contains 20 pencils. Five of the pencils are blue.
What is the probability that a pencil chosen at random is blue?
What is the probability that a pencil chosen at random is not blue?

Solution

Number of blue pencils is 5 out of 20.
Probability of choosing a blue pencil $= \frac{5}{20}$.
Number of pencils that are not blue is 15 out of 20.

Probability of choosing a non-blue pencil $= \frac{15}{20}$.

Notice that $\frac{15}{20} = \frac{20-5}{20}$

$= 1 - \frac{5}{20}$

So probability of choosing a non-blue pencil = 1 – probability of choosing a blue pencil.

> If the probability of an outcome is P, then the probability that the outcome will not happen is 1 – P.

The probability of an outcome is often denoted by P(outcome).

Example 1

The probability of drawing a 'diamond' card from a pack of 52 playing cards is $\frac{13}{52}$, that is $\frac{1}{4}$.

What is the probability of drawing a card that is not a diamond?

Solution

$\text{P(card is diamond)} = \frac{1}{4}$

$\text{P(card is not diamond)} = 1 - \frac{1}{4} = \frac{3}{4}$

Example 2

A bag contains blue, red and black marbles.
The probability of drawing a black marble is $\frac{3}{7}$.
If Bertha draws a marble from this bag, what is the probability that it is not black?

Solution

$\text{P(black)} = \frac{3}{7}$

$\text{P(not black)} = 1 - \frac{3}{7} = \frac{4}{7}$

This work on the probability of a single event is reasonably straightforward. You should be able to answer the questions in Exercise 7 without too much trouble.

EXERCISE 7

1. What is the probability that a card drawn from a shuffled pack of playing cards will be a king?

2. What is the probability that, when an unbiased die is rolled, it will turn up a prime number?

3. A bag contains 2 red balls, 3 white balls and 5 black balls. A ball is chosen at random. What is the probability that it is:
   a) red
   b) red or white
   c) neither red nor black

4. A bag contains 20 balls. 5 of the balls are blue.
   a) What is the probability that a ball chosen at random is blue?
   b) The probability of choosing a red ball is $\frac{1}{5}$. How many red balls are there in the bag?

5. The number of matches in each of 20 boxes of matches was counted. The results were as follows:

   39 43 42 40 41
   41 42 40 45 42
   43 40 39 41 39
   39 43 41 39 43

   a) Complete this frequency table.

Number of matches	39	40	41	42	43	44	45
Frequency	5	3					

   b) If one of these boxes is selected at random, what is the probability that it contains more than 40 matches?

6. 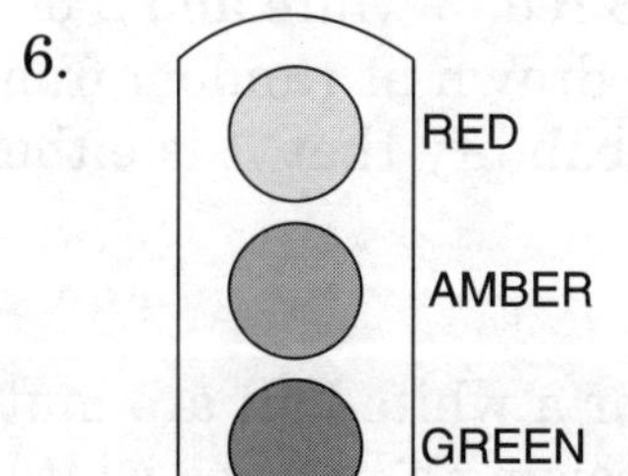

In any 50 second period, traffic lights at a road junction are
red for 25 seconds
green for 20 seconds and
amber for 5 seconds.

No two lights are 'on' at the same time.

A cyclist arrives at the junction.
What is the probability that the lights are:
   a) green
   b) red

Check your answers at the end of this module.

B Combining events – part 1

Mutually exclusive outcomes

Two or more events are said to be **mutually exclusive** if the happening of one event eliminates the happening of the other event(s). That is, at most one of these events can happen at any one time.

For example, when we toss a coin either 'head' or 'tail' may turn up but both events cannot happen in one throw. Similarly, in a single throw of a die 1, 2, 3, 4, 5 or 6 are mutually exclusive events.

If two events A and B are mutually exclusive and the probabilities of them occurring are P(A) and P(B) respectively, then the probability of either A or B happening is P(A) + P(B).

$$P(A \text{ or } B) = P(A) + P(B)$$

We can also apply the above rule to more than two events.

The probabilities of all possible outcomes add up to one.

Example 1

What is the probability of a die showing a 1 or a 6?

Solution

These are mutually exclusive events

$$P(1) = \frac{1}{6}$$

$$P(6) = \frac{1}{6}$$

$$P(1 \text{ or } 6) = P(1) + P(6)$$

$$= \frac{1}{6} + \frac{1}{6} = \frac{1}{3}$$

The probability of a die showing a 1 or a 6 is $\frac{1}{3}$.

Example 2

A bag contains 5 red, 3 white and 2 black balls, identical apart from colour. A ball is drawn at random from the bag.
What is the probability that it is either red or white?

Solution

Drawing a red or a white ball are mutually exclusive. Hence the probability of drawing a red or a white ball in a single draw is equal to the sum of the probability of drawing a red ball and the probability of drawing a white ball.

$$\text{The probability of drawing a red ball} = \frac{5}{10}$$

$$= \frac{1}{2}$$

$$\text{The probability of drawing a white ball} = \frac{3}{10}$$

$$\text{The probability of drawing a red or a white ball} = \frac{1}{2} + \frac{3}{10} = \frac{8}{10} = \frac{4}{5}$$

Example 3

The probabilities of three teams, P, Q, R, winning a soccer competition are $\frac{1}{4}, \frac{1}{8}, \frac{1}{10}$ respectively.

Calculate the probability that:

a) either P or Q will win
b) either P or Q or R will win
c) none of these teams will win

Solution

a) Since 'P will win' and 'Q will win' are mutually exclusive events the probability that either P or Q will win $= \frac{1}{4} + \frac{1}{8} = \frac{3}{8}$

b) The probability that either P or Q or R will win $= \frac{1}{4} + \frac{1}{8} + \frac{1}{10} = \frac{19}{40}$

c) The probability that none of these teams will win $= 1 - \frac{19}{40} = \frac{21}{40}$

EXERCISE 8

1. If an unbiased die is thrown, what is the probability that it will show a 6 or an odd number?

2. What is the probability of a card drawn from a pack being a 4 or a king?

3. A box contains 30 red, 20 blue and 10 green straws. What is the probability of a straw drawn at random being red or green?

4. There are a number of red, white and blue beads in a bag. One bead is drawn at random.
The probability of picking a red bead is $\frac{1}{3}$ and the probability of picking a blue bead is $\frac{1}{5}$.

   a) What is the probability of picking a bead which is red or blue?
   b) What is the probability of picking a white bead?

5. When three unbiased coins are tossed, the probability that they will show three 'heads' is 0.125. What is the probability that they will show at least one 'tail'?

Check your answers at the end of this module.

C Combining events – part 2

Independent events

An event A is said to be **independent** of another event B when the actual happening of A does not influence in any way the probability of the happening of B.

If A and B are two independent events then the probability of A and B occurring = probability of A occurring × probability of B occurring.

$$P(A \text{ and } B) = P(A) \times P(B)$$

For example, if a die and a coin are thrown, the score on the die has no effect on whether the coin comes down with head up or tail up.

Before doing a problem, you should be able to tell whether the events are mutually exclusive, independent or neither.

Example 1

If two dice are thrown, find the probability of getting two 4s.

Solution

The two throws are independent events.

Probability of getting a 4 on the first throw $= \frac{1}{6}$

Probability of getting a 4 on the second throw $= \frac{1}{6}$

Therefore, probability of getting two 4s $= \frac{1}{6} \times \frac{1}{6}$

$= \frac{1}{36}$

Example 2

Find the probability of scoring 18 when three dice are rolled.

Solution

To score 18, we must score a 6 on each die.

Probability of scoring a 6 on each throw $= \frac{1}{6}$

The probability of scoring three 6s is $= \frac{1}{6} \times \frac{1}{6} \times \frac{1}{6}$

$= \frac{1}{216}$

Example 3

The probability that a student is left-handed is 0.15.

a) What is the probability that the student is right-handed?
b) Two students are chosen at random. What is the probability that they are both left-handed?

Solution

a) 'The student is right-handed' is the same as 'The student is not left-handed'.

The probability that the student is right-handed
$= 1 -$ (the probability that the student is left-handed)
$= 1 - 0.15$
$= 0.85$

b) The probability that the first student is left-handed $= 0.15$.
The probability that the second student is left-handed $= 0.15$.

These events are independent, so the probability that both students are left-handed $= 0.15 \times 0.15$
$= 0.0225$

Diagrams

When an experiment consists of several independent or dependent trials, it is often convenient to display the various outcomes in a list or table or on a graph. Another way is to use a **tree diagram** in which the probabilities can be shown.

A probability tree shows all the possible events. It starts with a dot (the trunk of the tree). This dot represents the first event. From this dot 'branches' are drawn to represent all the possible outcomes from the event. The probability of each outcome is written on its 'branch'.

Suppose a die is thrown and a coin is tossed.
The first event of throwing a die can result in a 1, 2, 3, 4, 5 or 6, and we can show these as six branches starting from one point.
The probability of any one of these numbers appearing is $\frac{1}{6}$, and we can indicate this on the branches.

The first branch shows the appearance of 1 and the probability of it appearing.

The second branch shows the appearance of 2 and the probability of it appearing.

Similarly, the other branches illustrate the appearance of the other numbers and the corresponding possibilities.

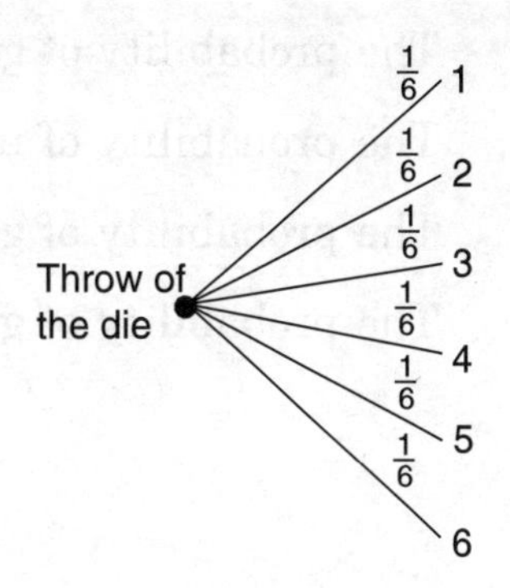

The second event of throwing the coin can follow any one of these outcomes. So we put a dot at the end of each branch to represent the next event of tossing the coin.

The tossing of the coin can result in a head or a tail – two possibilities.

Since the second event can follow any one of the first six outcomes, we can draw two more branches from each dot at the end of the first six branches. So, each branch will have two more branches.

The probability of a head appearing is $\frac{1}{2}$ and the probability of a tail appearing is also $\frac{1}{2}$.

We can put this on the branches. After adding the branches for the coin the tree diagram will look like the one given below.

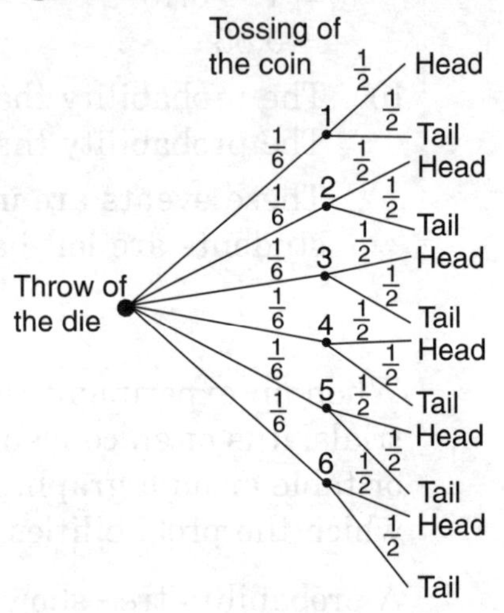

Suppose you want to know the probability of getting a number less than 5 and a 'head'.

The favourable outcomes are: (1, head), (2, head), (3, head), (4, head).

The number of favourable outcomes = 4.

The total number of outcomes = 12.

Therefore, the probability of getting a number less than 5 and a 'head' $= \frac{4}{12} = \frac{1}{3}$.

It is tedious to draw a tree with many branches like the one we drew earlier and so a condensed version is often used.

Since we are interested in the outcome 'less than 5', we draw the tree diagram as follows and label the branches with the appropriate probabilities.

The probability of getting a number less than 5 $= \frac{4}{6}$.

The probability of not getting a number less than 5 $= 1 - \frac{4}{6} = \frac{2}{6}$.

The probability of getting a head $= \frac{1}{2}$.

The probability of getting a tail $= \frac{1}{2}$.

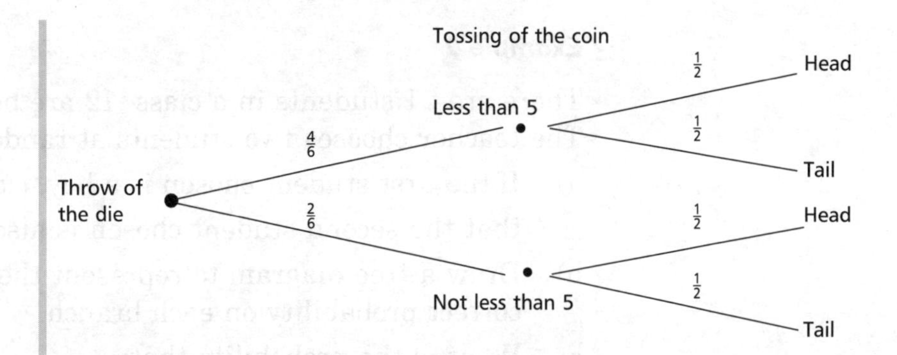

To find the probability of getting a number less than 5 and a head, first locate the 'less than 5' branch and then follow the branch 'head' that starts from the 'less than 5' branch. These are independent events.

Therefore, write down the probabilities on these branches and multiply them together.

The probability of getting a number less than 5 and a head $= \frac{4}{6} \times \frac{1}{2} = \frac{1}{3}$.

> When you are using a tree diagram, the probabilities on two consecutive branches must be multiplied to find the probability of a combined event. If the outcome of a combined event can be obtained in two or more mutually exclusive ways, it will be necessary to add the corresponding probabilities obtained by multiplying probabilities on the branches.

Example 1

Two unbiased coins are tossed together. Find the probability of getting:

a) two tails
b) one head and one tail

Solution

We can represent the two tosses by a tree diagram.

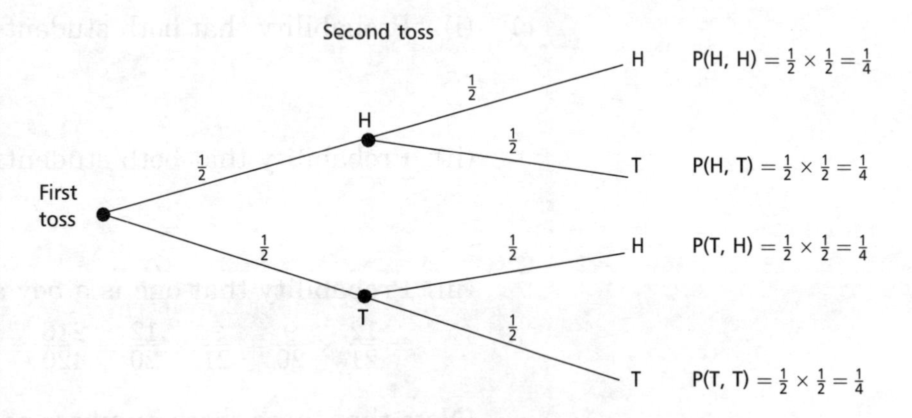

a) Probability of getting two tails $=$ P(T, T) $= \frac{1}{4}$.

b) Probability of getting one head and one tail $=$ P(H, T) $+$ P(T, H)

$$= \frac{1}{4} + \frac{1}{4} = \frac{1}{2}$$

Note that the sum of all the probabilities at each stage is 1.

Example 2

There are 21 students in a class. 12 are boys and 9 are girls. The teacher chooses two students at random.

a) If the first student chosen is a boy, explain why the probability that the second student chosen is also a boy is $\frac{11}{20}$.

b) Draw a tree diagram to represent the situation. Write the correct probability on each branch.

c) What is the probability that:

(i) both students are boys
(ii) both students are girls
(iii) one is a boy and one is a girl

d) The teacher chooses a third student at random. What is the probability that:

(i) all three students are boys
(ii) at least one of the three students is a girl

Solution

a) If the first student chosen is a boy, there are 20 students left and 11 of them are boys. Hence, the probability that the second student chosen is a boy $= \frac{11}{20}$.

b)

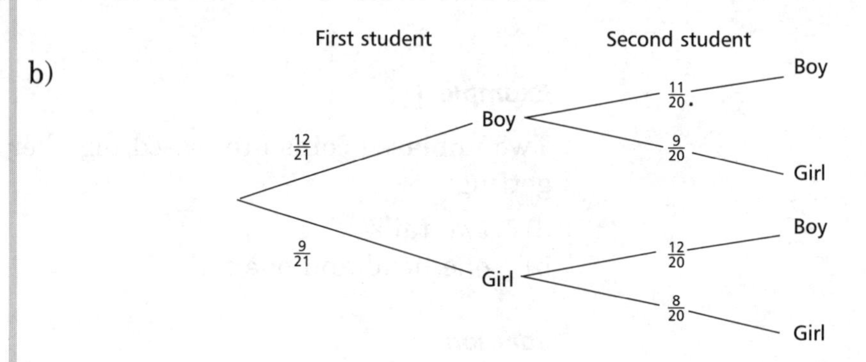

c) (i) Probability that both students are boys $= \text{P(B, B)}$

$$= \frac{12}{21} \times \frac{11}{20} = \frac{11}{35}$$

(ii) Probability that both students are girls $= \text{P(G, G)}$

$$= \frac{9}{21} \times \frac{8}{20} = \frac{6}{35}$$

(iii) Probability that one is a boy and one is a girl

$$= \frac{12}{21} \times \frac{9}{20} + \frac{9}{21} \times \frac{12}{20} = \frac{216}{420} = \frac{18}{35}$$

(Note that these three numbers add up to 1. This is because the outcome must be two boys, two girls or one boy and one girl – there are no other possibilities.)

d) (i) Probability that all three students are boys

$$= \text{P(B, B, B)} = \frac{12}{21} \times \frac{11}{20} \times \frac{10}{19} = \frac{22}{133}$$

(ii) 'At least one of the three students is a girl' is the same as 'The three students are not all boys'.

Probability that at least one of the three students is a girl
$= 1 -$ (probability that all three students are boys)
$= 1 - \frac{22}{133}$
$= \frac{111}{133}$

Example 3

A family with three children could have a girl first, followed by a boy, followed by another boy. This could be written GBB.

a) Use the letters B and G to make a list of *all* possible combinations in families with three children.

b) Assume that each of the combinations that you have listed is equally likely.
A family with three children is chosen at random. Find the probability that:
(i) it contains at least one girl
(ii) it contains two girls
(iii) the oldest and the youngest are of the same sex

Solution

a) The possible combinations are
GGG, GGB, GBG, GBB, BGG, BGB, BBG, BBB.
(Notice that there are two possibilities for the first child, two for the second child and two for the third child.
Total number of possibilities $= 2 \times 2 \times 2 = 8$.
The combinations could be obtained by drawing a tree diagram.)

b) (i) All the combinations contain at least one girl, except BBB.
Probability that family contains at least one girl $= \frac{7}{8}$.
(ii) Combinations containing exactly two girls are GGB, GBG and BGG.
Probability that family contains exactly two girls $= \frac{3}{8}$.
(iii) Combinations in which the oldest and the youngest are of the same sex are GGG, GBG, BGB, BBB.
Probability that the oldest and the youngest children are of the same sex $= \frac{4}{8} = \frac{1}{2}$.

This work demands concentration and I hope you have understood the examples. See whether you can obtain the correct answers to the following questions.

EXERCISE 9

1. An unbiased coin is tossed twice. What is the probability of the two tosses giving the same result?

2. A bag contains 8 blue marbles and 2 red marbles. Two marbles are drawn at random. What is the probability of getting:
   a) two red marbles
   b) one red marble and one blue marble
   c) two blue marbles

3. A bag contains 12 beads. Five are red and the rest are white. Two beads are drawn at random. Find the probability that:
   a) both beads are red
   b) both beads are white

4. Mahmoud enjoys flying his kite. On any given day, the probability that there is a good wind is $\frac{3}{4}$.
   If there is a good wind, the probability that the kite will fly is $\frac{5}{8}$.
   If there is not a good wind, the probability that the kite will fly is $\frac{1}{16}$.

   a) (i)

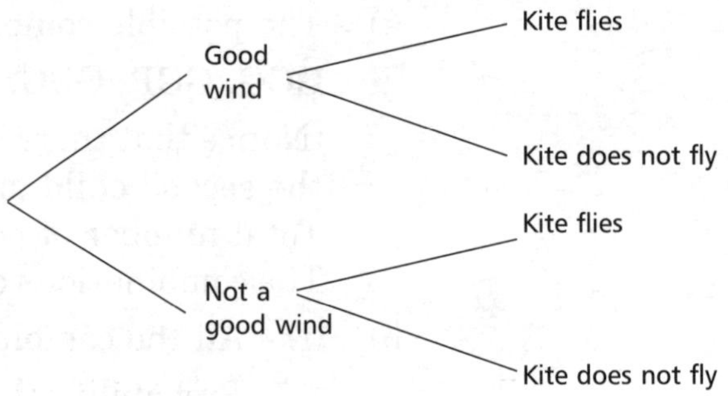

   Good wind — Kite flies / Kite does not fly
   Not a good wind — Kite flies / Kite does not fly

   Copy the given tree diagram.
   Write on your diagram the probability for each branch.

   (ii) What is the probability of a good wind *and* the kite flying?

   (iii) Find the probability that, whatever the wind, the kite does *not* fly.

   b) If the kite flies, the probability that it sticks in a tree is $\frac{1}{2}$. Calculate the probability that, whatever the wind, the kite sticks in a tree.

I hope that you were able to answer all these questions correctly.

Check your answers at the end of this module.

Summary

In this unit on probability you have learnt that:

- the probability of a favourable outcome $= \dfrac{\text{the number of favourable outcomes}}{\text{the number of possible equally likely outcomes}}$
- if the probability of an outcome is P then the probability that the outcome will not happen is $1 - P$
- $P(A \text{ or } B) = P(A) + P(B)$
- $P(A \text{ and } B) = P(A) \times P(B)$ and you learnt how to draw a tree diagram to help you determine the probability of combined events.

The next unit in the module focuses on geometrical transformations and the use of vectors.

Check your progress

1. There are 240 pupils at Denton School. The pie chart (which is not to scale) shows how they travel to school.

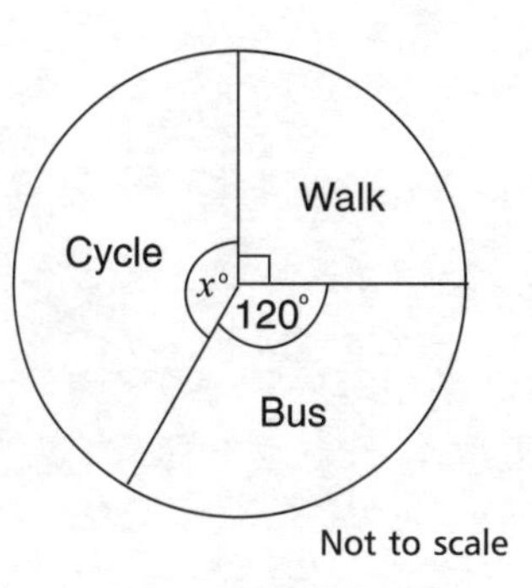

   a) Calculate the value of x.
   b) How many pupils walk to school?
   c) What is the probability that a pupil, chosen at random, travels to school by bus?

2. Sian has three cards, two of them black and one red. She places them side by side, in random order on a table.
   One possible arrangement is red, black, black.
   a) Write down all the possible arrangements.
   b) Find the probability that the two black cards are next to one another.
   Give your answer as a fraction.

3. A die has the shape of a tetrahedron.
   The four faces are numbered 1, 2, 3 and 4.
   The die is thrown on the table. The probabilities of each of the four faces finishing flat on the table are as shown.

Face	1	2	3	4
Probability	$\frac{2}{9}$	$\frac{1}{3}$	$\frac{5}{18}$	$\frac{1}{6}$

   a) Fill in the four empty boxes with the probabilities changed to fractions with a common denominator.
   b) Which face is most likely to finish flat on the table?
   c) Find the sum of the four probabilities.
   d) What is the probability that face 3 does *not* finish flat on the table?

4. The diagram below shows all the possible outcomes when two fair dice are thrown.

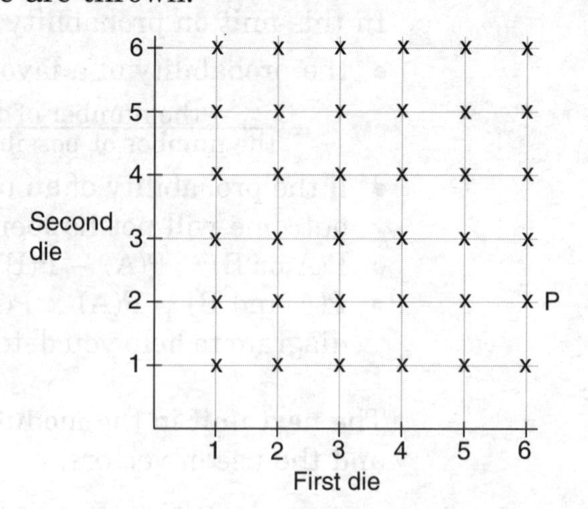

a) Explain clearly what outcome is represented by the cross P.
b) Copy the diagram and ring the crosses representing all those outcomes with a total score on the two dice of 8.
c) Find the probability, as a fraction in its lowest terms, that:
   (i) the two dice will show a total score of 8
   (ii) the two dice will show the same score as each other

5. The probability that Jane will pass her English examination is $\frac{3}{4}$, and the probability that she will pass her mathematics examination is $\frac{4}{5}$.

What is the probability that:

a) she will not pass in English
b) she will pass in both subjects

6. 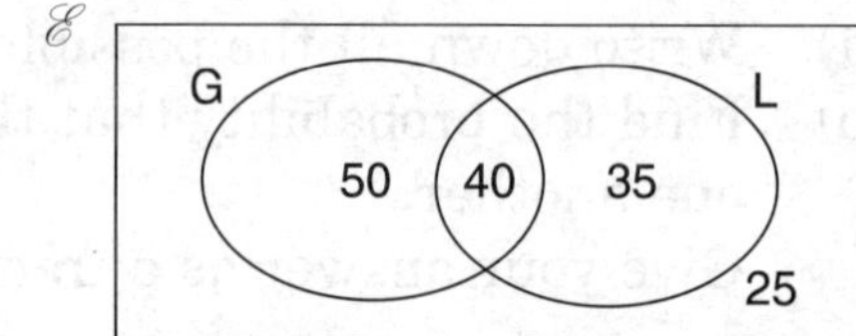

ℰ = (students in an international school)
G = (girls)
L = (students who speak more than one language)

The Venn diagram shows the number of students in each subset, in a school of 150 students.

a) (i) How many girls speak only one language?
   (ii) How many boys are there in the school?
b) Give your answers to the following questions as fractions in their lowest terms.
   (i) A student is selected at random. What is the probability that this student speaks more than one language?
   (ii) A girl is selected at random. What is the probability that she speaks more than one language?

(iii) A student who speaks more than one language is selected at random. What is the probability that this student is a girl?

(iv) Two students are selected at random. What is the probability that they are both boys?

7. In this question, give all your probabilities as fractions in their lowest terms.

a) Six chairs are placed in a row.

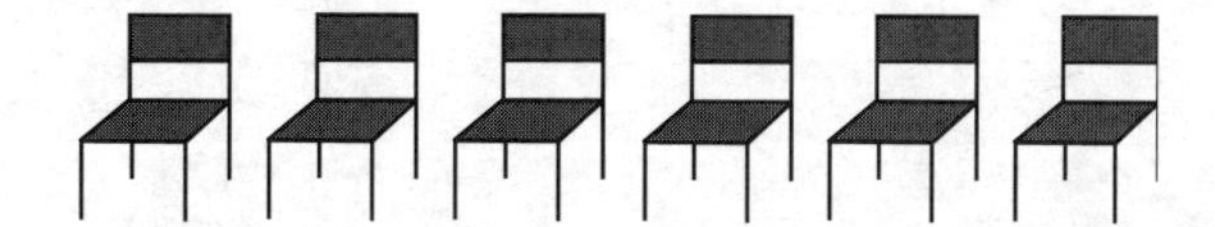

Alain is equally likely to sit on any one of the chairs.

(i) What is the probability that Alain sits on one of the end chairs?

(ii) After Alain has sat down, Bernard chooses any chair at random. What is the probability that Bernard sits next to Alain:

(a) if Alain is sitting at an end

(b) if Alain is not sitting at an end

(iii) Copy the tree diagram and write the probabilities on each branch.

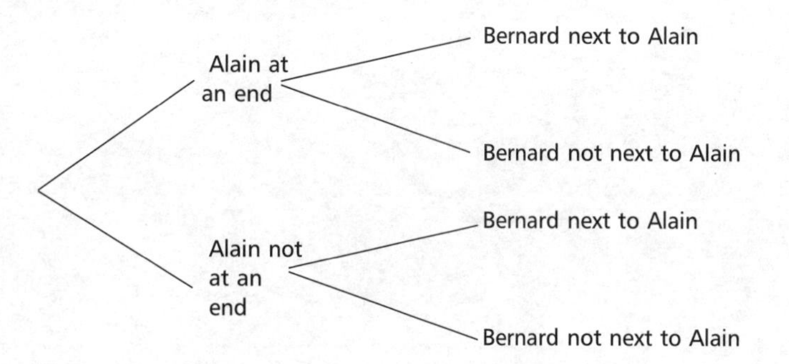

(iv) Find the probability that Bernard sits next to Alain, wherever Alain sits.

b) The six chairs are placed in a circle and Alain and Bernard sit down.

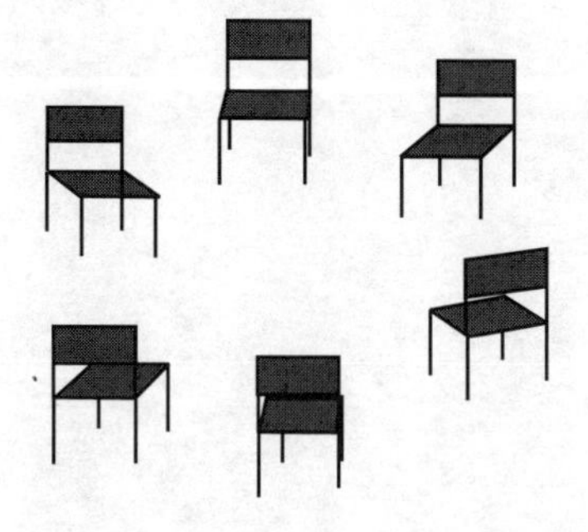

What is the probability that Bernard sits next to Alain?

c) There are n chairs in a circle and Alain and Bernard sit down. The probability that Bernard sits next to Alain is $\frac{1}{4}$. Find the value of n.

You have now completed the work on statistics and probability. We hope that you enjoyed it and that you were able to answer the questions.

Check your answers at the end of this module.

Unit 3
Geometrical Transformations and Vectors

In this unit we will be looking at the movement of geometrical figures in a plane. You will learn how to describe these movements mathematically, in some cases with the help of vectors.

This unit is divided into three sections:

Section	Title	Time
A	Simple transformations	4 hours
B	Vectors	4 hours
C	Further transformations	3 hours

By the end of this unit, you should be able to:

- reflect simple plane figures in a given line
- rotate simple plane figures about a point
- construct translations and enlargements of simple plane figures
- identify and give descriptions of transformations
- describe a translation by means of a vector
- add and subtract vectors
- calculate the magnitude of a vector
- use position vectors
- use the sum and difference of two vectors to express given vectors in terms of two coplanar vectors
- identify and use sheer and stretch transformations.

A Simple transformations

Look at the figures given below. What do you see in these figures?

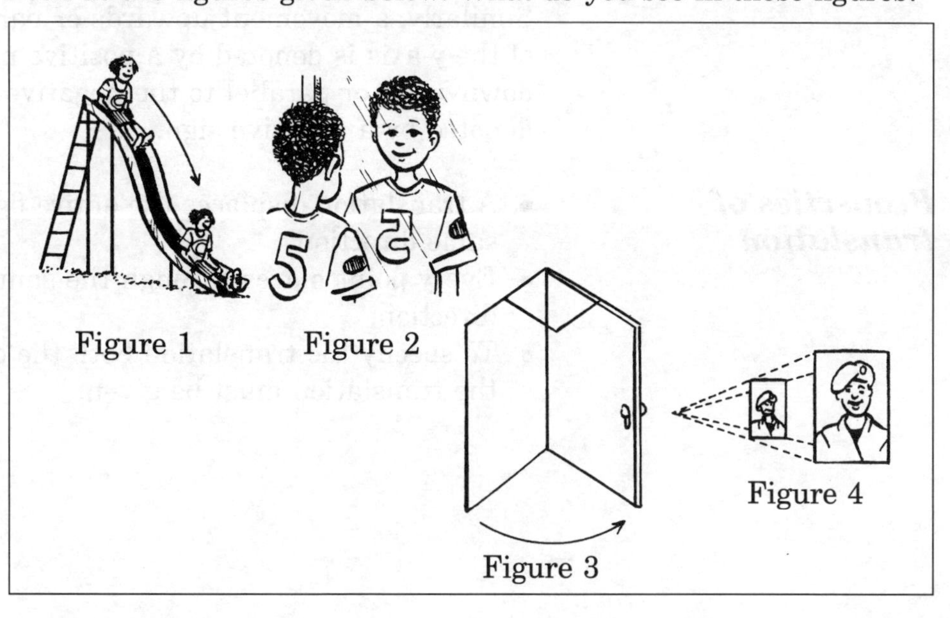
Figure 1 Figure 2 Figure 3 Figure 4

In Figure 1 the child has moved a certain distance, d, in a certain direction along the line. Figure 2 shows the reflection of a girl in a mirror. In Figure 3 you see a door after it has been turned through an angle of 90°. The last figure shows the enlargement of a picture.

All the above movements are called **transformations**. A transformation is an operation that can change the position and/or the shape and/or size of an object.

Different transformations produce different effects and each of these transformations has a particular name.

In geometry if a point or a set of points (a figure) moves from one place to another, it is said to have undergone a transformation. That is why this is sometimes called motion geometry.

In a transformation one point or a figure is **mapped** onto another. In this context 'mapped' means the same as transferred. The position of a point or a set of points (a figure) after transformation is often called the **image** of the original one under the mapping or the transformation.

Any transformation which leaves a figure unchanged in size and shape is called an **isometric transformation.** Figures which are of the same shape and size can be called isometric, but this word is not very often used. Usually they are called congruent figures.

The three main isometric transformations are translation, reflection and rotation. Points which are in the same position before and after a transformation are called **invariant points** or **fixed points**.

Translation

In general, a **translation** (slide) is a motion of a specified distance and along a straight line without any accompanying twisting or turning.

A movement parallel to the positive direction of the x-axis or to the right is denoted by a positive sign, and a movement parallel to the negative direction of the x-axis or to the left is written with a negative.

Similarly, a movement upwards or parallel to the positive direction of the y-axis is denoted by a positive number, and a movement downwards or parallel to the negative direction of the y-axis is denoted by a negative sign.

Properties of translation

- A translation displaces the entire figure the same distance in the same direction.
- Every point moves through the same distance in the same direction.
- To specify the translation both the distance and the direction of the translation must be given.

- We can name the translation of an entire object by specifying the translation undergone by one point in the figure.

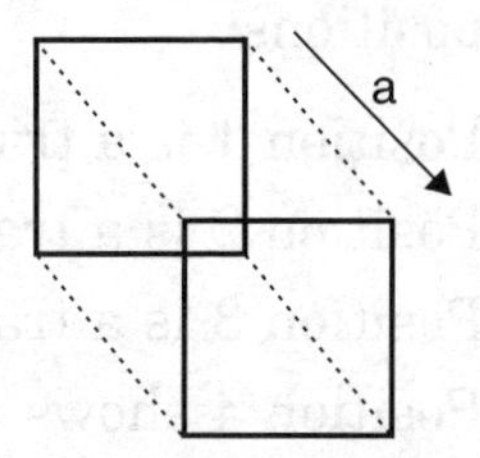

A translation of *a* units in the direction of the arrow

- No part of the figure is invariant; that is, no point stays in the original position.
- The original figure and the image are **directly congruent.** This means that the size, the shape, *and* the orientation (the way the figure appears) of the object remain unchanged. Study the following explanation to understand the difference between direct congruence and indirect congruence.

The orientation of Figure 1 is in the order ABC. One has to move in the anticlockwise sense to go from A to B, B to C and then back to A.

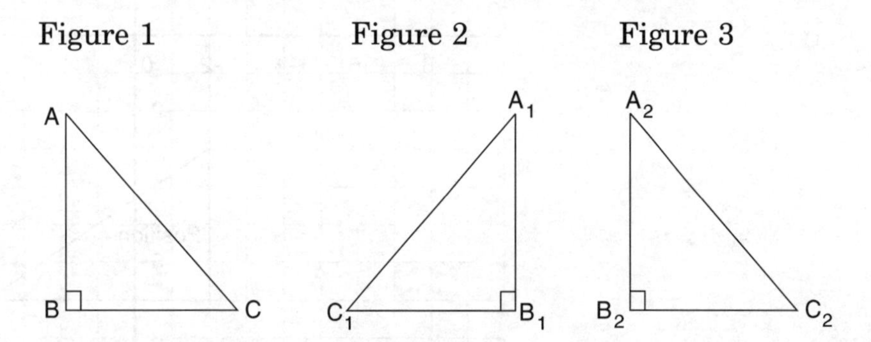

In Figure 2 the orientation is in the order $A_1B_1C_1$.
The movement from A_1 to B_1, B_1 to C_1 and then back to A_1 is in the clockwise sense.

In Figure 3 the orientation of $A_2B_2C_2$ is in the anticlockwise sense.

All three figures are congruent.
While the figure ABC and the figure $A_1B_1C_1$ are indirectly congruent, figures ABC and $A_2B_2C_2$ are directly congruent.

Example and solution

In the given diagram below, the triangle T is translated to 5 positions:

Position 1 is a translation 7 units to the right parallel to the x-axis.
Position 2 is a translation 4 units upward parallel to the y-axis.
Position 3 is a translation 7 units to the left parallel to the x-axis.
Position 4 shows a translation of 8 units downward parallel to the y-axis.

Position 5 shows two movements 7 units to the right parallel to the x-axis and 8 units downwards parallel to the y-axis.

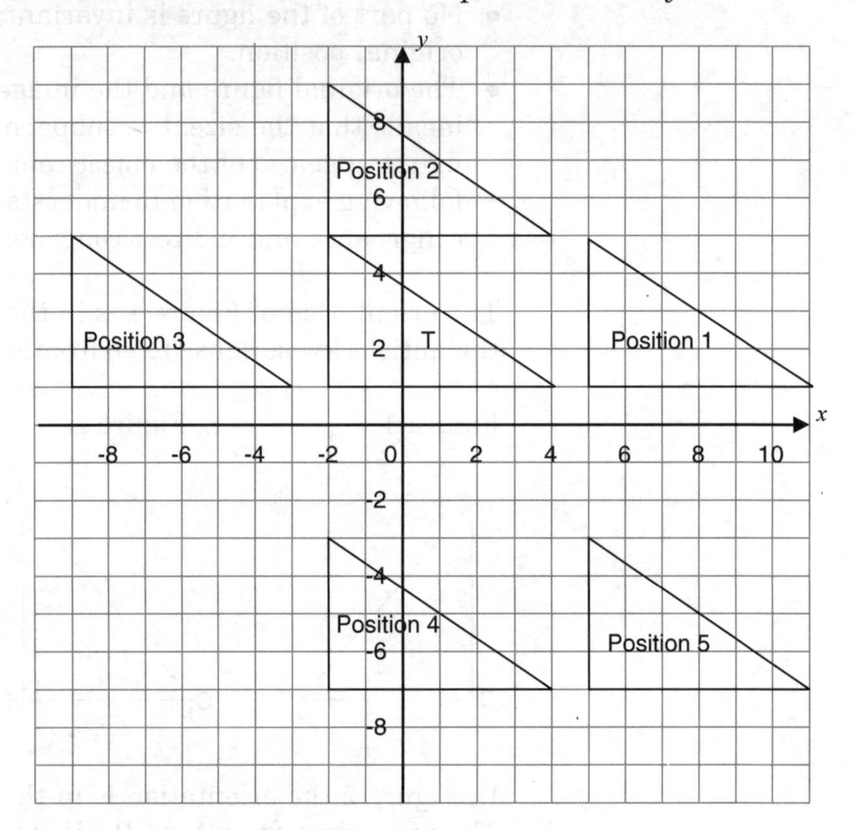

EXERCISE 10

1. Draw sketches to illustrate the following translations:
   a) a square is translated 6 cm to the left
   b) a triangle is translated 5 cm to the right

2. Translate the triangle ABC given in the diagram:
   a) 3 units to the right
   b) 3 units to the left
   c) 3 units upwards
   d) 3 units downwards

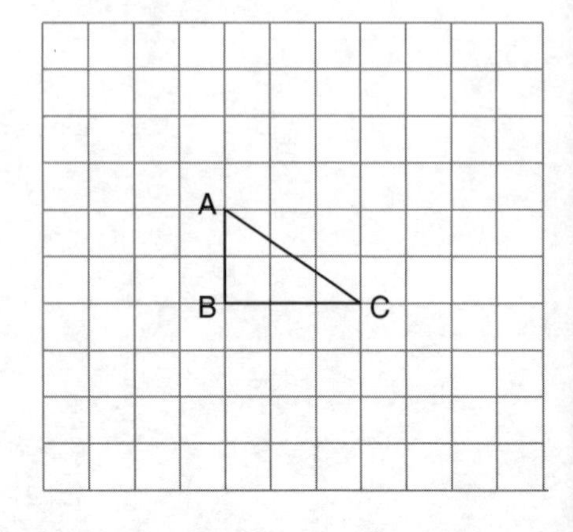

3. On squared paper, draw x and y axes and mark the points A(3, 5), B(2, 1) and C(−1, 4).
   a) Mark and label the triangle $A_1B_1C_1$ which is the image of triangle ABC under the translation of 2 units parallel to the positive direction of the x-axis and 3 units parallel to the negative direction of the y-axis.
   b) Mark and label the triangle $A_2B_2C_2$ which is the image of triangle ABC under the translation of 4 units to the right parallel to the x-axis and 1 unit upwards parallel to the y-axis.

Check your answers at the end of this module.

Reflection

A **reflection** in a line l is a motion that pairs each point P of a shape with a point P′ in such a way that l is the perpendicular bisector of the line PP′ as long as P is not on line l. If P is on l, then P′ is the same point as P.

> Reflection is a type of transformation in which a point and its image have a symmetry about a line.

In the diagram below, P′ is the image of P by reflection in the line AB, and P is the image of P′.
PO = OP′ and angle AOP = 90°.

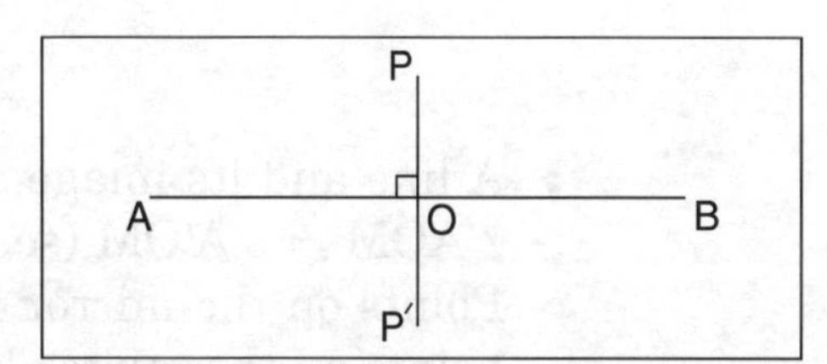

Reflection maps any point P onto a unique image P′ and every image arises from just one point P. This means that there is a one-to-one correspondence between each point and its image.

Reflection in a real mirror only works from one side.
If you stand behind a mirror you will not produce an image.

> A mathematical mirror is double-sided and reflects both ways.

Properties of reflection

- A point (P) and its image (P′) after reflection in line l are equidistant from the line (see Figure 1 below).

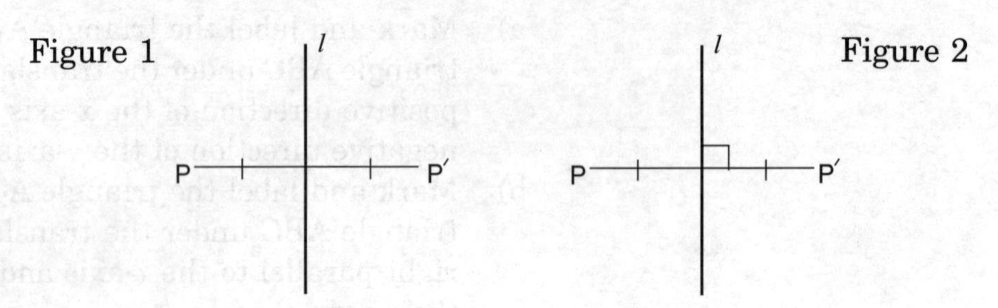

- The mirror line bisects the line joining a point and its image at right angles (see Figure 2 above).
- A line segment and its image are equal in length: AB = A′B′ (see Figure 3 below).

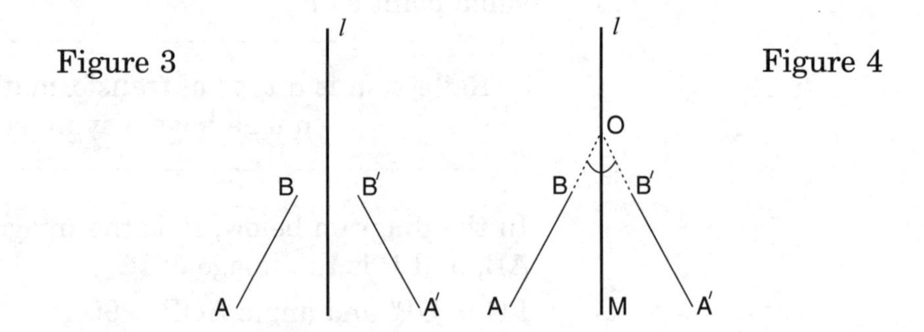

- A line and its image are equally inclined to the mirror line: $\angle$ AOM = $\angle$ A′OM (see Figure 4 above).
- Points on the mirror line are their own images. In other words, points on the mirror line are invariant (do not vary) (see Figure 5 below).
- A figure and its image are indirectly congruent (see Figure 6 below).

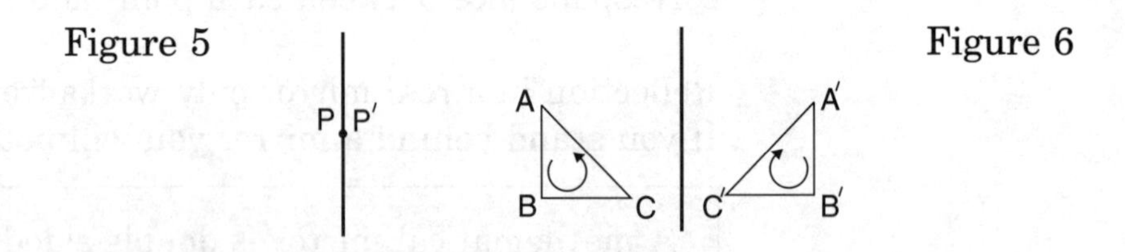

Triangles ABC and A′B′C′ are congruent, but their orientations are opposite. Triangle ABC is in the anticlockwise sense and triangle A′B′C′ is in the clockwise sense.

Example 1

Reflect the triangle ABC in the mirror line (indicated by a dashed line).

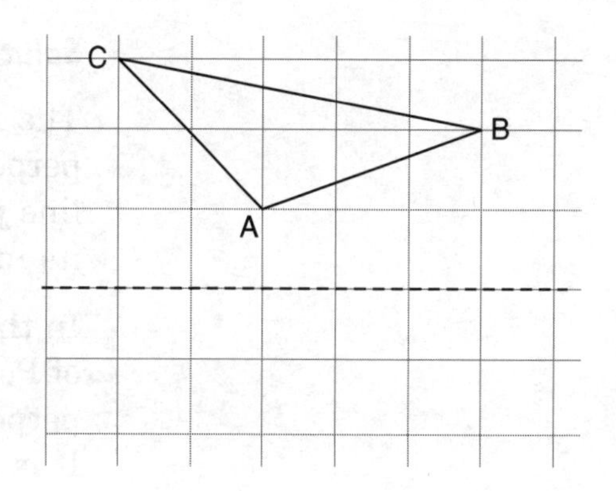

Solution

The mirror line is one of the grid lines. This makes it easy to reflect any point. You simply count the squares from the point to the mirror line, and the reflection is the same distance the other side of the mirror line.

In the diagram, A is 1 unit from the mirror line, so its image A′ is also 1 unit from the mirror line. Point B is 2 units from the mirror line, so its image B′ is also 2 units from the mirror line. Similarly for C and its image C′.

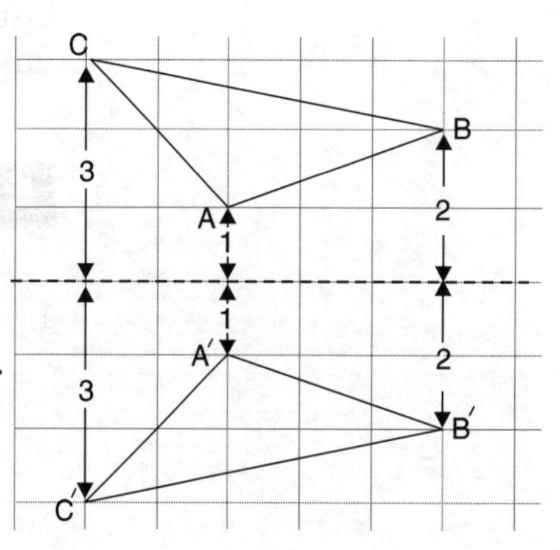

The reflection of a straight line is a straight line. So, to obtain the reflection of triangle ABC, join A′ to B′, B′ to C′ and C′ to A′.

Check: You can check your result by using tracing paper. Mark two points X and Y on the mirror line. Trace the triangle ABC and the points X and Y.
Turn the tracing paper over and place it so that X and Y are exactly on top of their original positions.
If your diagram is correct, the traced triangle ABC will be exactly on top of triangle A′B′C′.

Example 2

The diagram shows a triangle and its reflection in a mirror line.

Draw the mirror line.

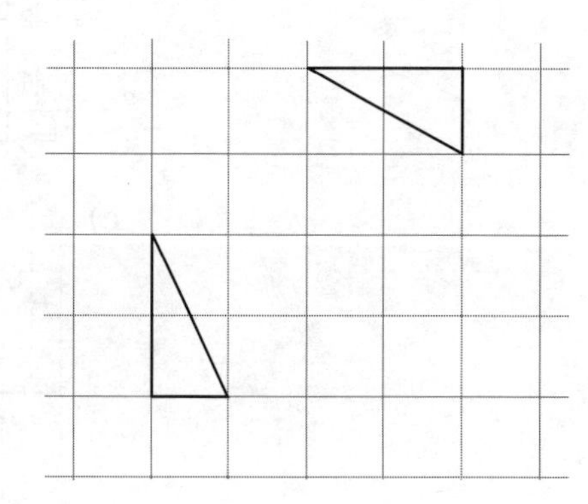

Solution

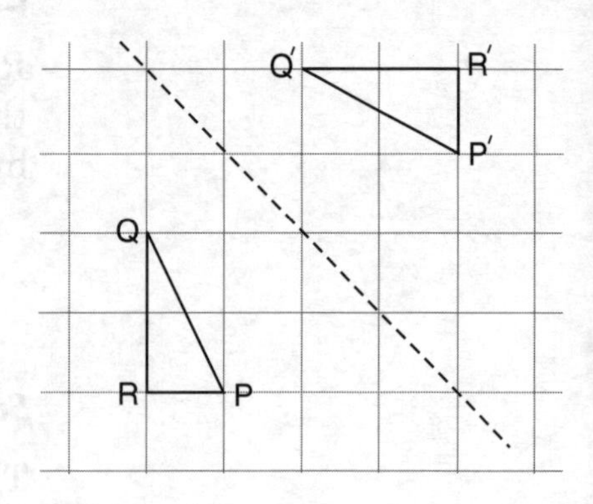

The mirror line is the perpendicular bisector of the line joining any point and its image.

In the diagram, P′ is the image of P. The mirror line is the perpendicular bisector of PP′. It is shown as a dashed line in the diagram.

Note: The same result is obtained by drawing the perpendicular bisector of QQ′ or RR′. (If that is not the case, you have made a mistake!)

EXERCISE 11

1. Copy the shapes and the mirror lines (indicated by dashed lines) onto squared paper and draw the image of each object.

a) 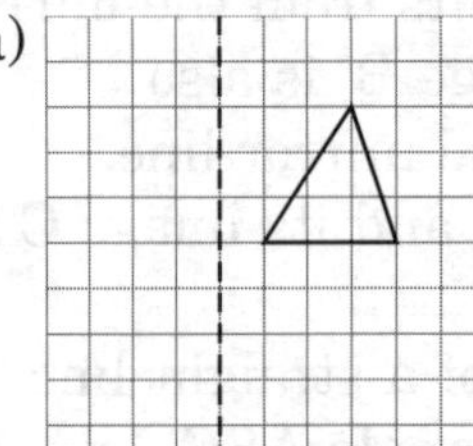

b) 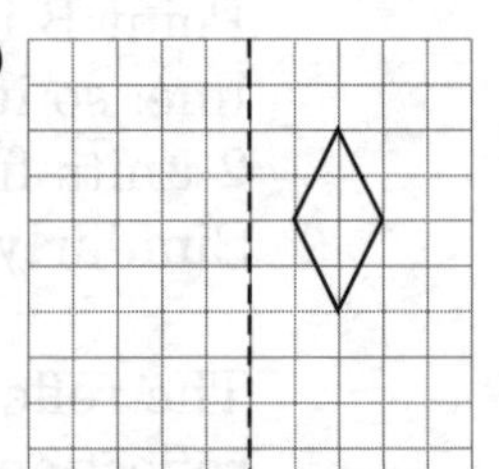

c) 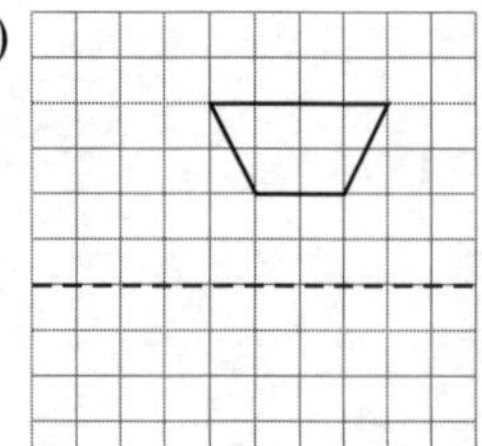

2. Copy the shapes given in parts a) to d).
   In each diagram, the shapes are reflections of one another.
   Draw the mirror line in each diagram.

a) 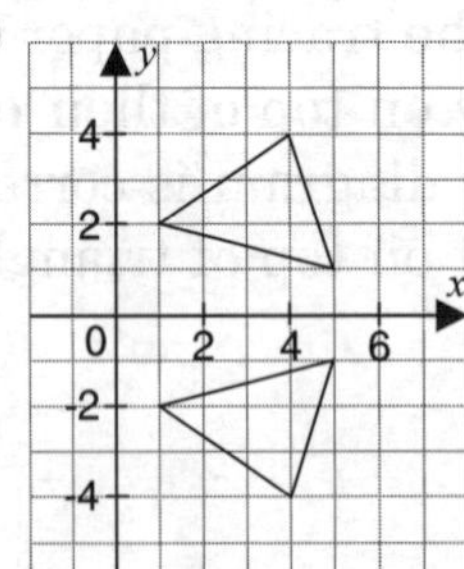

b) 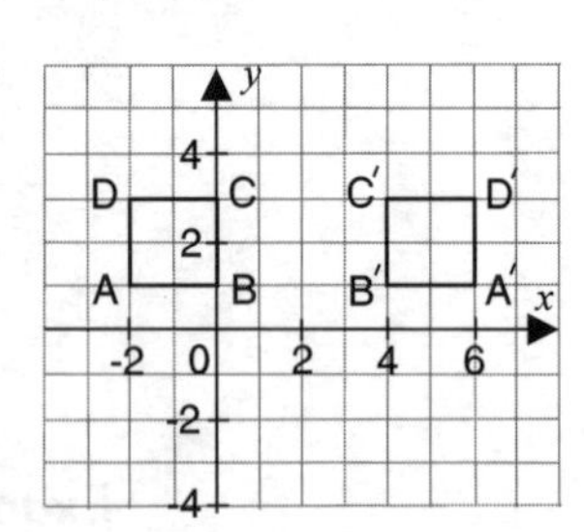

c) 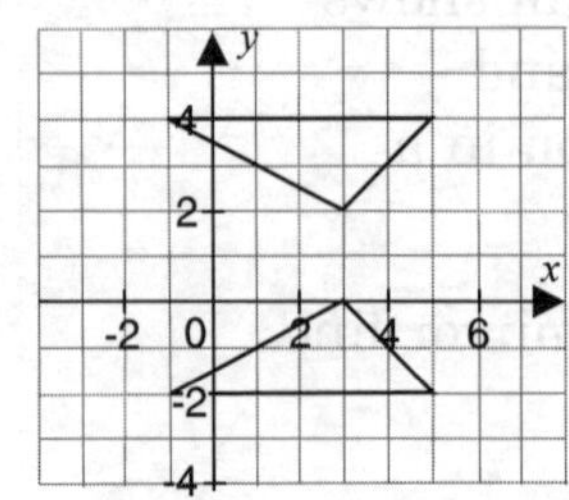

d) 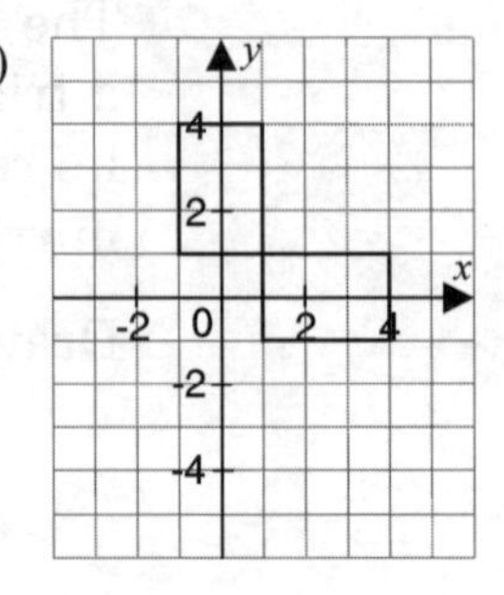

Check your answers at the end of this module.

Rotation

Rotation occurs when an object is turned about a given point. A bicycle wheel can be rotated about its centre. A square kite could be rotated about one of its corners. You can probably think of many other examples.

> A rotation (or a turn) is a transformation in which the shape is turned clockwise or anticlockwise through a particular angle about a fixed point.

The fixed point is called the **centre of rotation** and the angle through which the shape is rotated is called the **angle of rotation**.

Properties of rotation

- An anticlockwise rotation is taken as positive, and a clockwise rotation is said to be negative.
- A rotation through 180° is called a half-turn, and a rotation through 90° is called a quarter-turn.

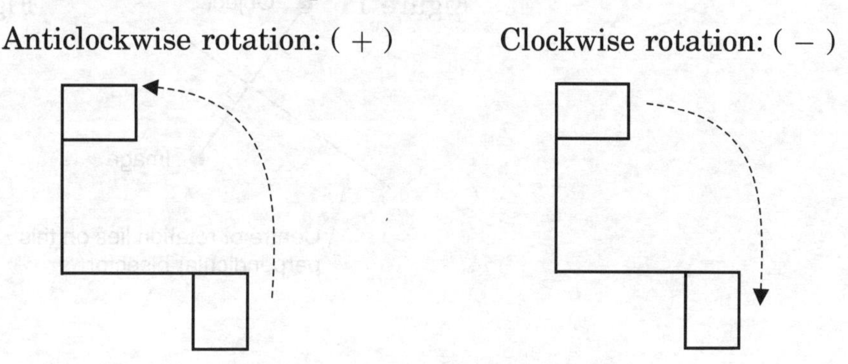

- A point and its image are equidistant from the centre of rotation.

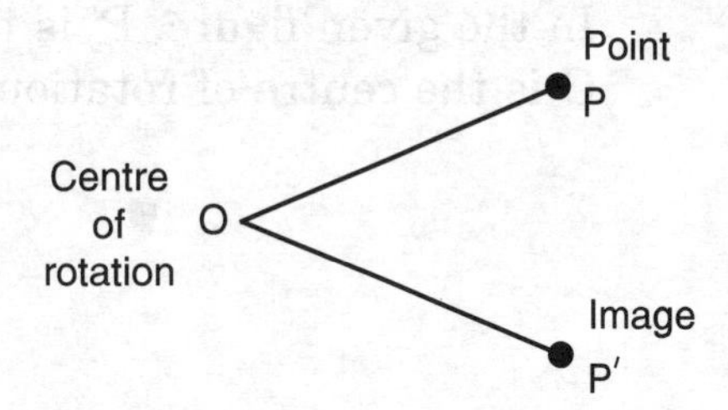

- Each point of a shape moves along the arc of a circle whose centre is the centre of rotation. So all the circles are concentric.

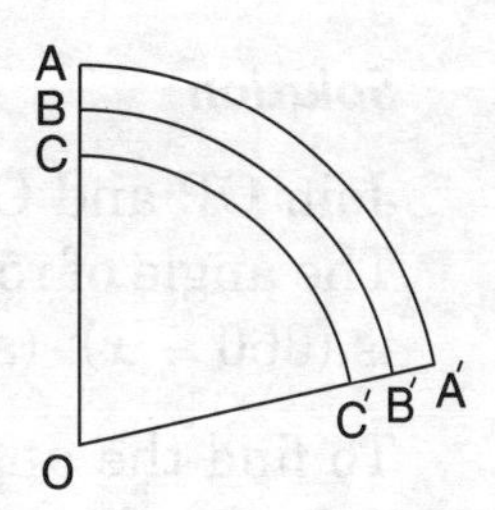

- Only the centre of rotation is invariant.

- The shape and its image are directly congruent.

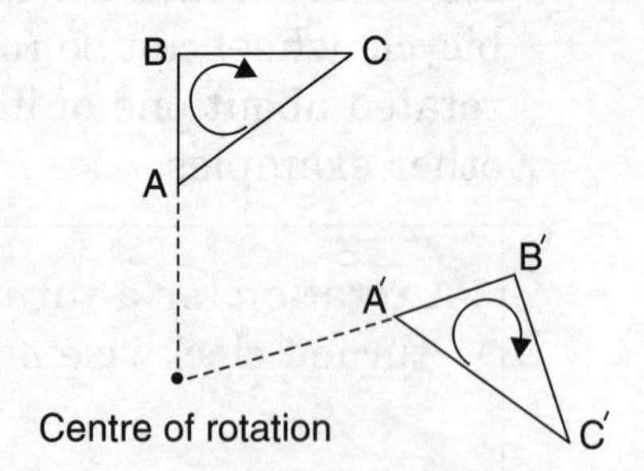

Triangle ABC and its image A′B′C′ are directly congruent.
ABC is oriented clockwise; A′B′C′ is also oriented clockwise.

- The perpendicular bisector of the line joining a point and its image passes through the centre of rotation (Figure 1).

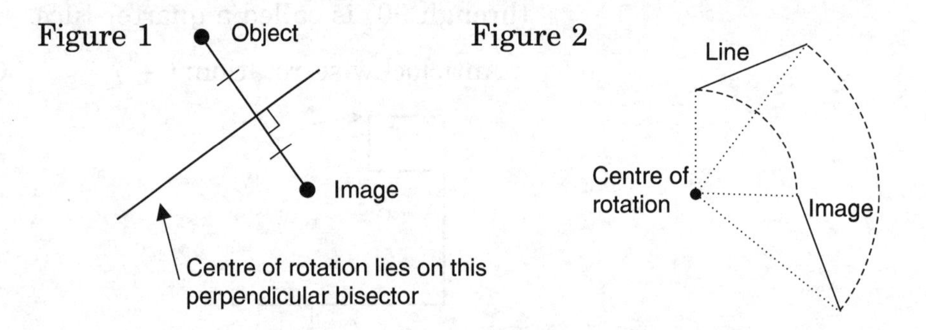

- A line segment and its image are equal in length (Figure 2).

Finding the angle of rotation

Example

In the given figure, P′ is the image of P after a rotation.
O is the centre of rotation. Find the angle of rotation.

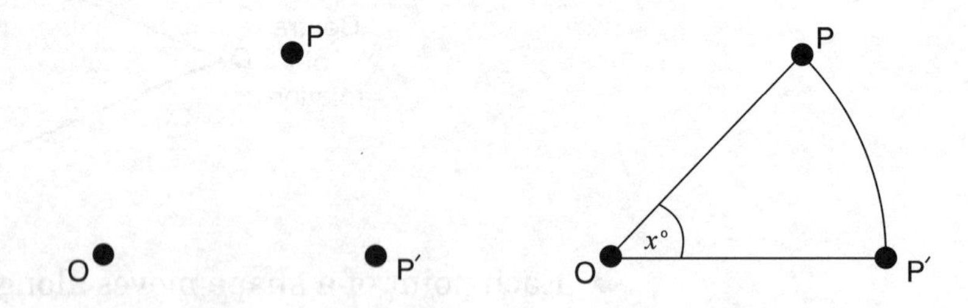

Solution

Join OP and OP′. Let angle POP′ be $x°$.
The angle of rotation is $-x°$ (clockwise sense) or the angle of rotation is $(360 - x)°$ (anticlockwise sense).

To find the angle of rotation, choose any point on the given figure. Join the chosen point to the centre of rotation. Find the image of the chosen point after rotation. Join it to the centre of rotation. Measure the angle between the two lines. Fix the sign of the angle according to the direction of the rotation.

Finding the centre of rotation

Example

A′B′ is the image of a line AB after rotation about the centre of rotation O. Find the position of the centre of rotation O.

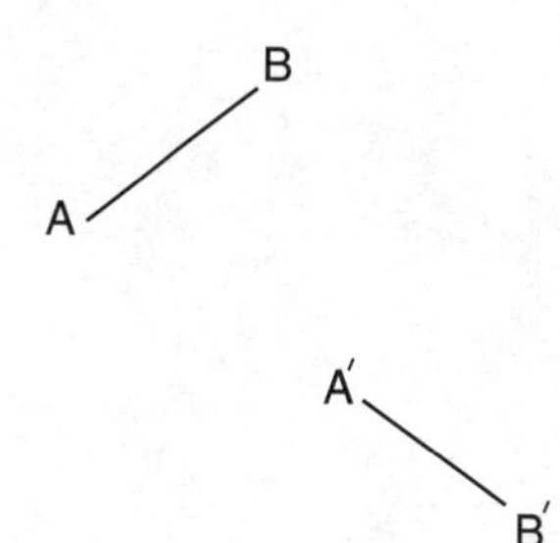

Solution

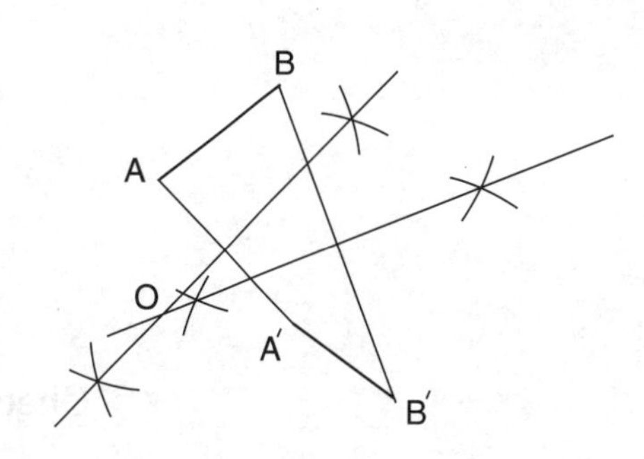

A and A′ are equidistant from the centre of rotation. The centre of rotation lies on the perpendicular bisector of AA′. Join AA′. Draw the perpendicular bisector of AA′ using a compass and a ruler.

B and B′ are also equidistant from the centre of rotation. The centre of rotation lies on the perpendicular bisector of BB′ as well. Join BB′. Draw the perpendicular bisector of BB′.

The point of intersection of the two perpendicular bisectors is O. O is the centre of the rotation.

You're probably finding that the work is getting more difficult. Don't worry! Some practice will help to increase your confidence.

EXERCISE 12

1. Copy the diagrams in questions a) to c).
   Find the images of the given triangles under the rotations described.

   a) Centre of rotation (0,0); angle of rotation 90° anticlockwise.
   b) Centre of rotation (3,1); angle of rotation 180°.
   c) Centre of rotation (−1,0); angle of rotation 180°.

remember that if there is no sign it is '+'

a) 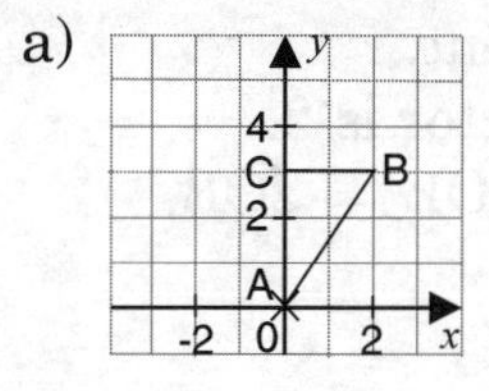

b) 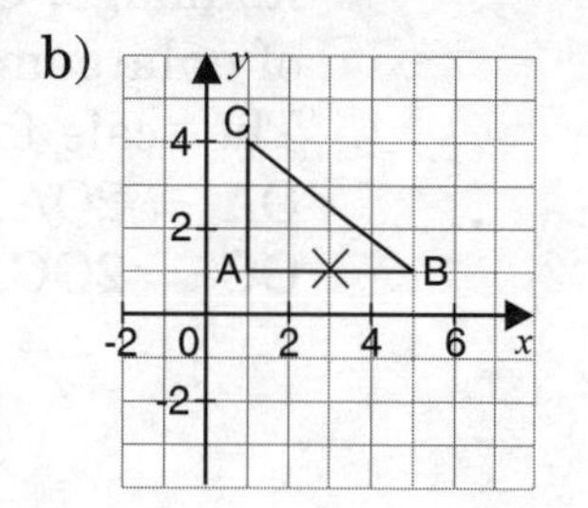

c) 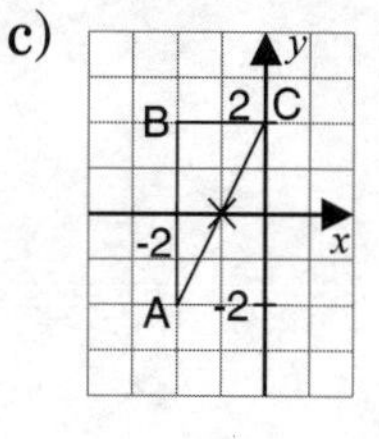

2. For each of the diagrams below, give the angle of the rotation which maps triangle ABC onto triangle A′B′C′ .

a)

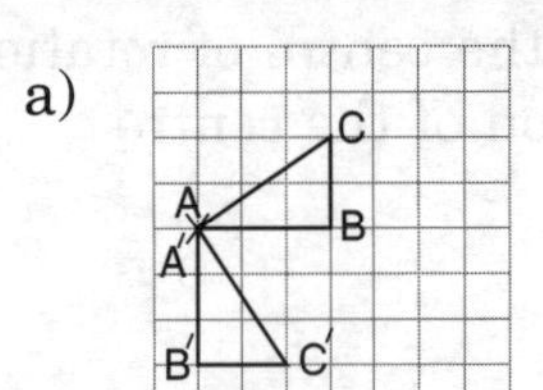

b)

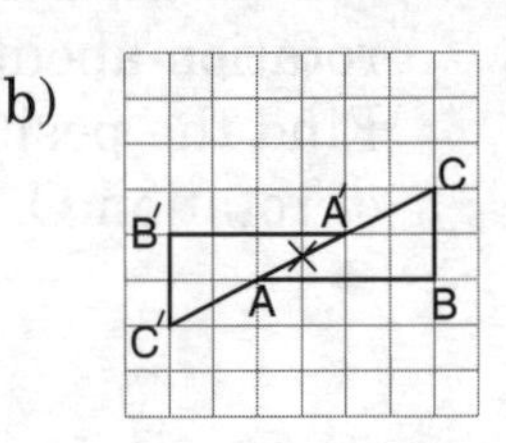

c)

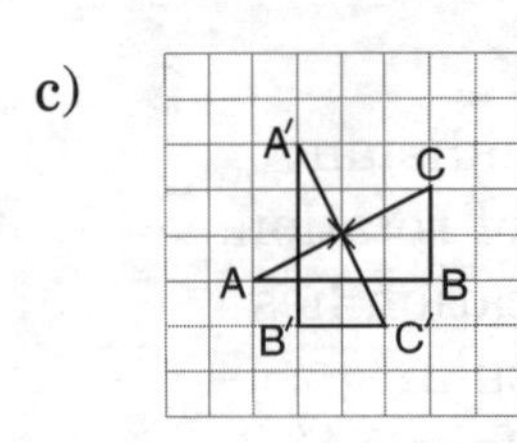

d)

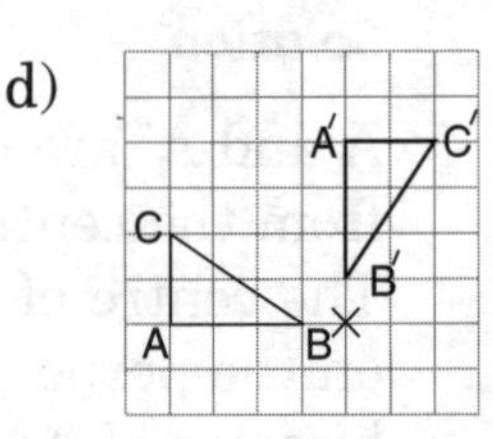

Check your answers at the end of this module.

Enlargement

The transformations translation, reflection and rotation moved the figure and perhaps turned it over to produce the image, but its shape and size are not changed. The **enlargement** keeps the shape but alters the size.

Enlargement is a transformation in which each point of a figure is mapped along a straight line drawn from a fixed point. The fixed point is called the **centre of enlargement**. The distance each point moves is a certain number times its distance from the centre of enlargement. The ratio, the length of a line segment : the length of its image, is equal to the scale factor.

Properties of enlargement

- The centre of enlargement can be anywhere, including a point inside the figure on the boundary of the figure.
- If the centre of enlargement is O and the scale factor is k, then each point A is mapped onto point A′ such that OA′ $= k$OA and O, A and A′ are collinear. If the scale factor is negative, then OA′ is in the opposite direction to OA and OA′ $= -k$OA.

A scale factor that is less than 1 produces an image that is smaller than the object, although the word enlargement is often still used to describe this transformation.

ABC is the figure. A′B′C′ is its image. O is the centre of enlargement.
The scale factor is 2.
OA′ = 2OA, OB′ = 2OB,
OC′ = 2OC.

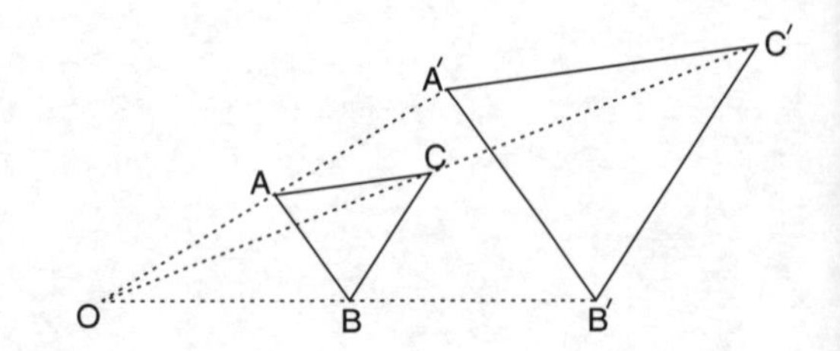

ABC is the figure. A′B′C′ is its image. O is the centre of enlargement.
The scale factor is $\frac{1}{2}$.

$OA' = \frac{1}{2}OA$, $OB' = \frac{1}{2}OB$,
$OC' = \frac{1}{2}OC$.

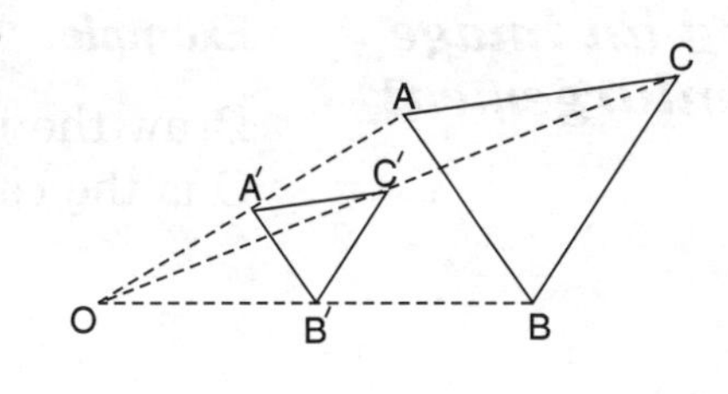

ABC is the figure. A′B′C′ is its image. O is the centre of enlargement.
The scale factor is -2.

$OA' = -2OA$, $OB' = -2OB$,
$OC' = -2OC$.

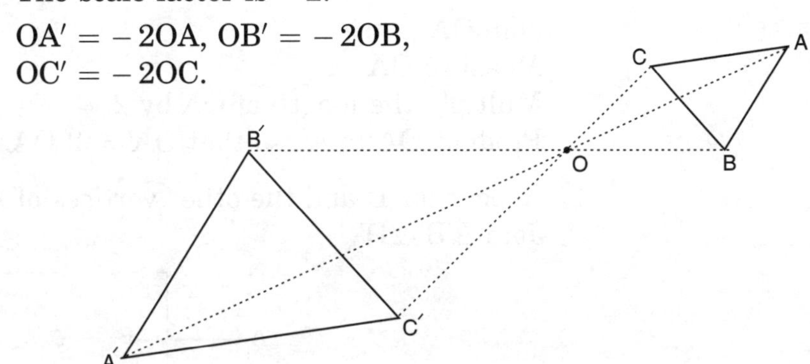

- A figure and its image are similar (not congruent) with sides in the ratio $1 : k$
- The areas of the figure and its image are in the ratio $1 : k^2$
- Angles and orientation of a figure are invariant.

Finding the centre of enlargement and scale factor

Example

The figure shows an object ABCD and its image A′B′C′D′ under an enlargement. Find the centre of enlargement and the scale factor.

Solution

Join the point A and its image A′. Extend AA′ in both directions. Similarly draw and extend BB′, CC′ and DD′.

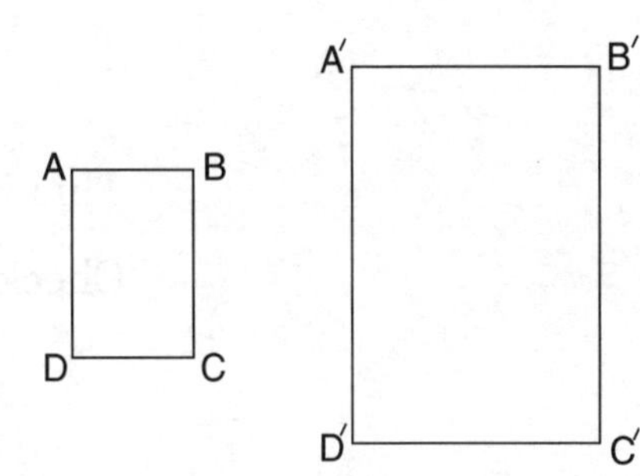

The point of intersection of these lines is the centre of enlargement, O. Measure OA′ and OA. The ratio OA′ : OA gives the scale factor.

> If a point and its image are on opposite sides of the centre of enlargement, then the scale factor is negative.

Drawing an image under enlargement

Example

Draw the image of the given rectangle ABCD.
O is the centre of enlargement and the scale factor is 2.

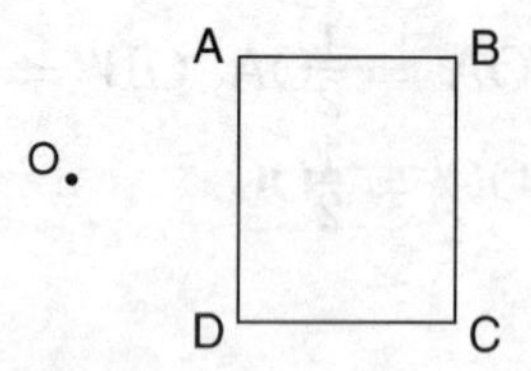

Solution

Join OA.
Measure OA.
Multiply the length of OA by 2.
Produce OA to A′, so that OA′ = 2 OA.

Repeat for B and the other vertices of ABCD.
Join A′B′C′D′.

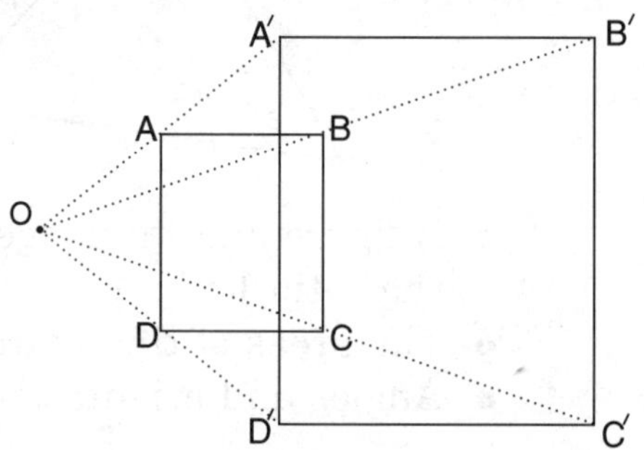

This is the image of ABCD under the given enlargement.

Check: In the completed diagram, A′B′ should be parallel to AB, B′C′ should be parallel to BC, C′D′ should be parallel to CD and D′A′ should be parallel to DA.
Also, the length of A′B′ should be twice the length of AB, etc.

See whether you have the right ideas about enlargement by answering the following questions.

EXERCISE 13

1. Draw the image of triangle ABC under an enlargement with scale factor 2 and centre of enlargement P(2, 1).

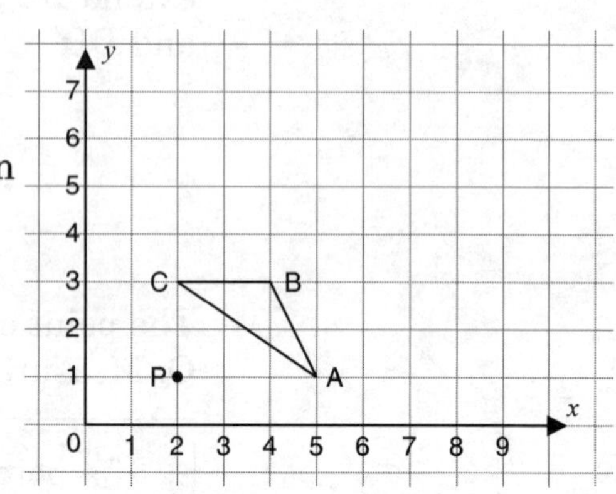

2. Draw the image of triangle DEF under an enlargement with scale factor -3 and centre of enlargement P(2, 0).

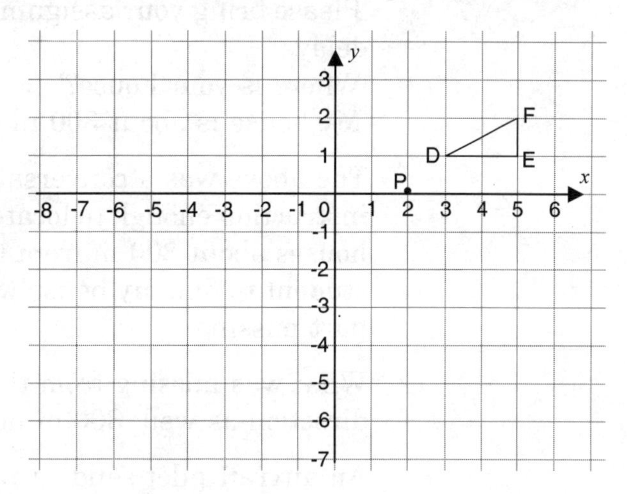

3. In the diagram, triangle G′H′I′ is the image of triangle GHI under an enlargement.

   Find the scale factor of the enlargement and the coordinates of the centre of enlargement.

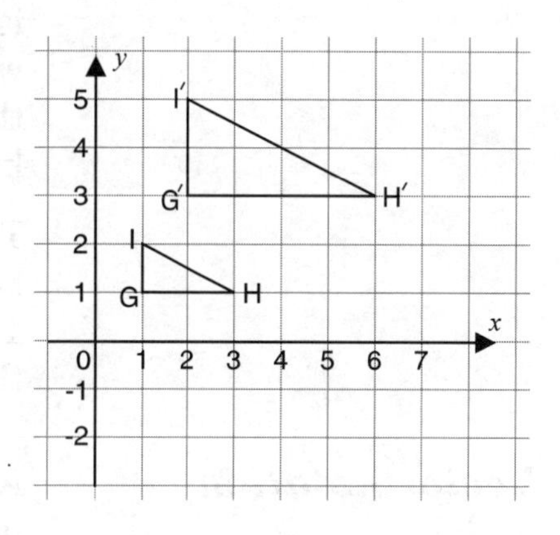

4. In the diagram, rectangle P′Q′R′S′ is the image of rectangle PQRS under an enlargement.

   Find the scale factor of the enlargement and the coordinates of the centre of enlargement.

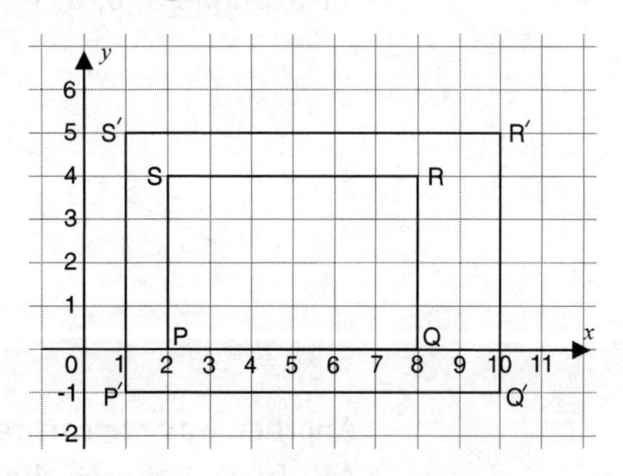

Check your answers at the end of this module.

B Vectors

'Please bring your assignments to my house. This way you can avoid delays.'
'Where is your house?'
'My house is about 300 m from the University.'

The above was a conversation between a student and me. Was my instruction enough to locate my house? No, there may be a number of houses about 300 m from the University. It will not be easy for the student to find my house with the information I gave her. There is a part missing.

What was missing from the instruction? I should have given the direction as well: 300 m north, or east or something like that.

An aircraft pilot sends a radio message to the airport he is approaching to ask for a report on the wind. 'Wind speed 40 km/h', he is told. Is this enough? Certainly not. If the wind is behind him, he will get there earlier than expected. If it is ahead, he will be late. If there is a cross wind, he will be blown off course. So the pilot needs to know both the speed *and* direction of the wind.

There is a special name for these two-part quantities, which include both magnitude and direction. They are called **vector** quantities. The magnitude part tells us how far, how much, how fast, etc., and the other part tells us in what direction.

Vector notation

A vector can be represented on paper by an arrow of a certain length proportional to the magnitude and pointing in the direction of the vector. In other words, a vector can be represented by a directed line segment.

We often denote a vector by a single small letter in bold italic type, for example $\boldsymbol{a}$, $\boldsymbol{b}$, $\boldsymbol{u}$, $\boldsymbol{v}$. As it is not easy to copy this in your books, you can write a vector as a small letter with a wavy line underneath it, for example $\underset{\sim}{a}$, $\underset{\sim}{b}$, $\underset{\sim}{u}$, $\underset{\sim}{v}$.

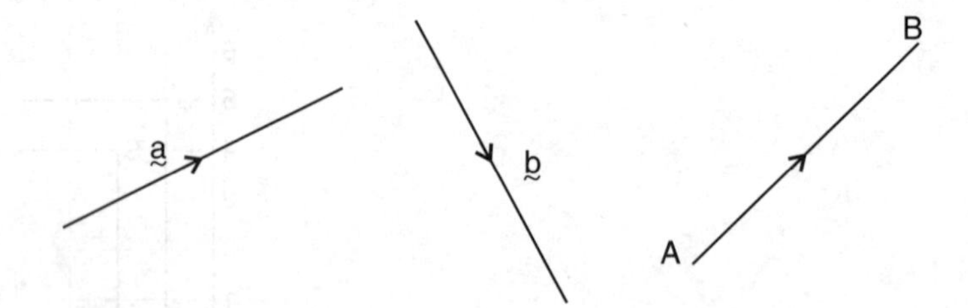

Another way we can represent a vector is by a named line such as AB. In such a case the vector is denoted by **AB** or $\overrightarrow{AB}$. The order of letters is important as it gives the direction. $\overrightarrow{AB}$ is not the same as $\overrightarrow{BA}$. $\overrightarrow{AB}$ and $\overrightarrow{BA}$ have the same magnitude but opposite directions.

Number pair notation

We can represent vectors by number pair notation.

The line PQ represents the displacement from P to Q.
This also represents the translation of P to Q.

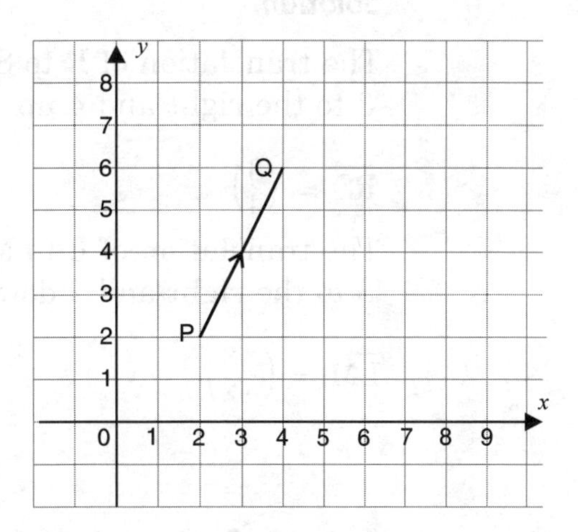

The displacement is 2 units in the positive x-direction, followed by 4 units in the positive y-direction.

This can be written as the ordered pair $\binom{2}{4}$.

The top number shows the horizontal movement (parallel to the x-axis) to the right or left. The bottom number shows the vertical movement (parallel to the y-axis) upwards or downwards.

Movements in the positive x-direction (that is, movement to the right) and movements in the positive y-direction (that is, upward movements) are indicated by positive numbers. Normally the positive sign is not shown.

Movements in the negative x-direction (movement to the left) and movements in the negative y-direction (downward movements) are indicated by negative numbers.

Two units to the right is written 2. Four units upwards is written 4. On the other hand, 2 units to the left is written -2 and 4 units downwards is -4. No movement is denoted by ( ).

Thus, we can write $\mathbf{PQ} = \overrightarrow{PQ} = \binom{2}{4}$

This number pair notation is called a **column vector**.

Example 1

Express the vectors $\overrightarrow{RS}$ and $\overrightarrow{LM}$ in number pair form.

Solution

The translation of R to S is '3 to the right and 4 up'.

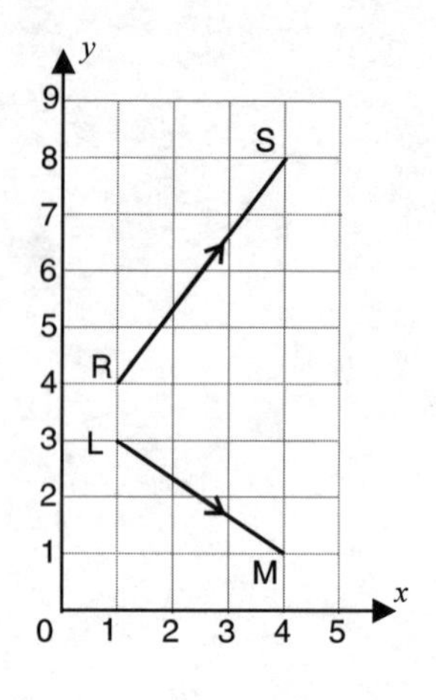

$\overrightarrow{RS} = \binom{3}{4}$

The translation of L to M is '3 to the right and 2 down'.

$\overrightarrow{LM} = \binom{3}{-2}$

Example 2

Draw the column vectors $\binom{1}{3}$ and $\binom{-2}{-4}$.

Solution

Start at any convenient point, say A, and move '1 to the right and 3 up' to B.

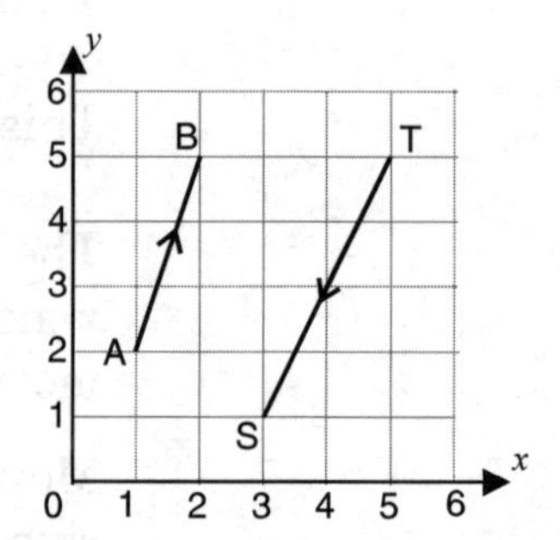

Thus, $\overrightarrow{AB} = \binom{1}{3}$.

Start at any point, say T, and move '2 to the left followed by 4 down' to S.

Thus, $\overrightarrow{TS} = \binom{-2}{-4}$.

Translation described by a vector

Column vectors can be used to describe translations. For example, triangle ABC is translated to triangle A′B′C′. All the points on the triangle ABC have moved 2 units to the right and 3 units upwards.

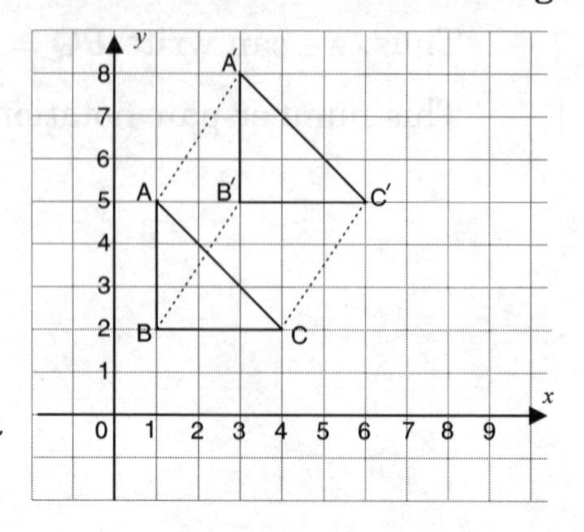

For example, $\overrightarrow{AA'} = \binom{2}{3}$, $\overrightarrow{BB'} = \binom{2}{3}$ and $\overrightarrow{CC'} = \binom{2}{3}$.

Therefore, the column vector for the above translation is $\binom{2}{3}$.

Example

Rectangle R is translated to rectangle S.
Find the column vector for the translation.

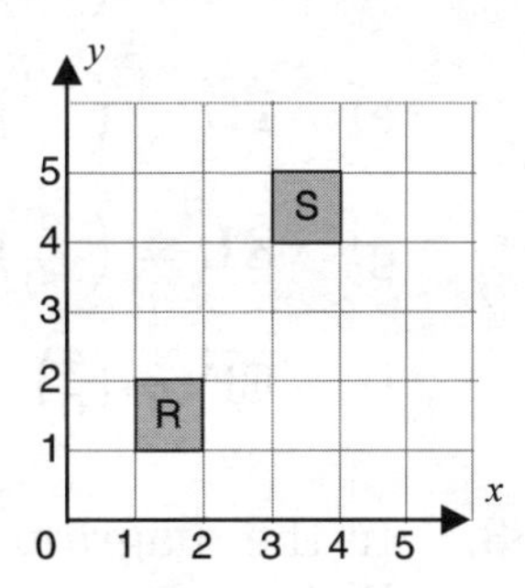

Solution

Each point on the rectangle moves 2 units to the right and 3 units upwards. To make it easier, just consider one vertex of the rectangle and its image.

Thus, the column vector is $\binom{2}{3}$.

Now try the questions in Exercise 14.

EXERCISE 14

1. Write column vectors for each of the vectors shown on the diagram given below.

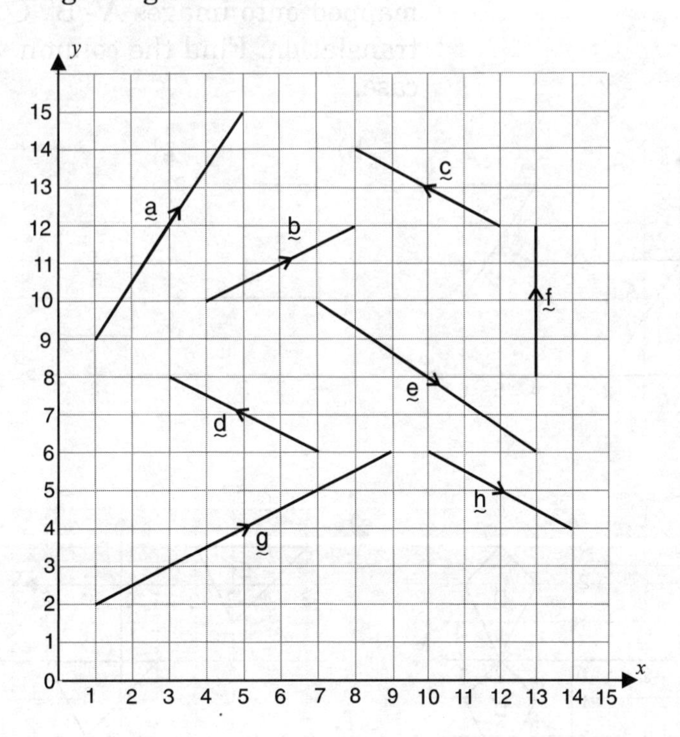

2. On squared paper, represent these vectors:

a) $\overrightarrow{AB} = \begin{pmatrix} 5 \\ 2 \end{pmatrix}$ b) $\overrightarrow{CD} = \begin{pmatrix} 2 \\ 2 \end{pmatrix}$

c) $\overrightarrow{PQ} = \begin{pmatrix} -1 \\ 3 \end{pmatrix}$ d) $\overrightarrow{RS} = \begin{pmatrix} 0 \\ 3 \end{pmatrix}$

e) $\overrightarrow{TU} = \begin{pmatrix} -2 \\ 0 \end{pmatrix}$ f) $\overrightarrow{MN} = \begin{pmatrix} -2 \\ -4 \end{pmatrix}$

g) $\overrightarrow{KL} = \begin{pmatrix} 0 \\ -5 \end{pmatrix}$ h) $\overrightarrow{VW} = \begin{pmatrix} -3 \\ -3 \end{pmatrix}$

i) $\overrightarrow{EF} = \begin{pmatrix} 4 \\ 0 \end{pmatrix}$ j) $\overrightarrow{JI} = \begin{pmatrix} -3 \\ -2 \end{pmatrix}$

3. In the diagram below, ABCD is a parallelogram. Write column vectors for the following:

a) $\overrightarrow{AB}$ and $\overrightarrow{DC}$

b) $\overrightarrow{BC}$ and $\overrightarrow{AD}$

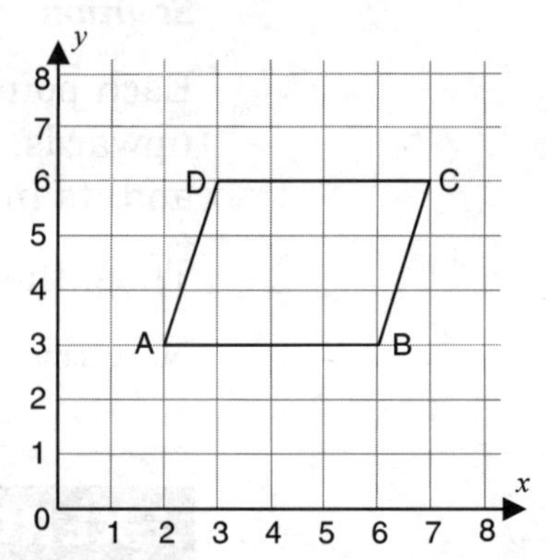

What can you say about the two pairs of vectors?

4. In the diagrams shown below, shapes A, B, C, D, E and F are mapped onto images A′, B′, C′, D′, E′ and F′ respectively by a translation. Find the column vector for the translation in each case.

a)
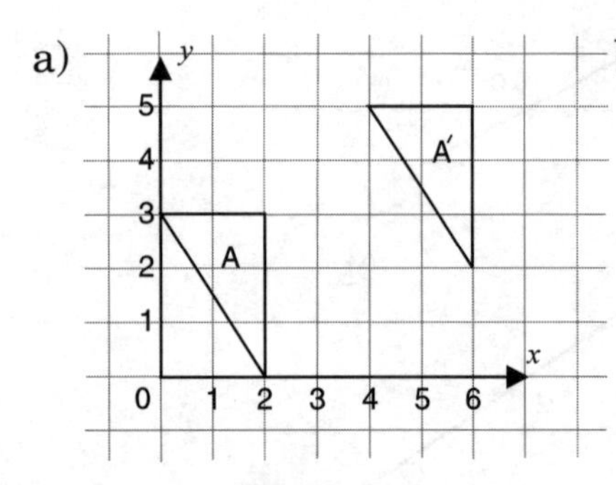

b)
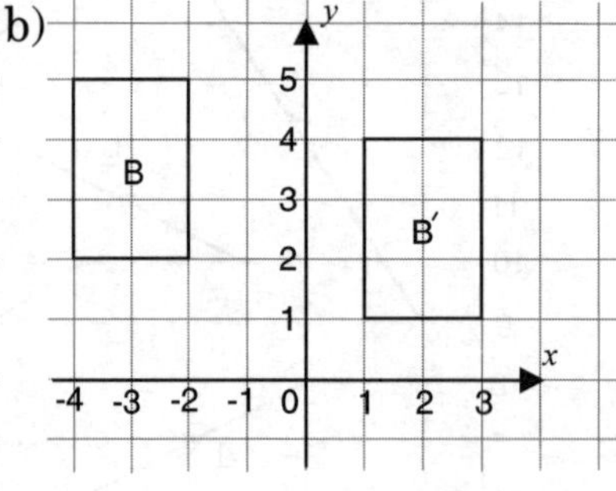

c)
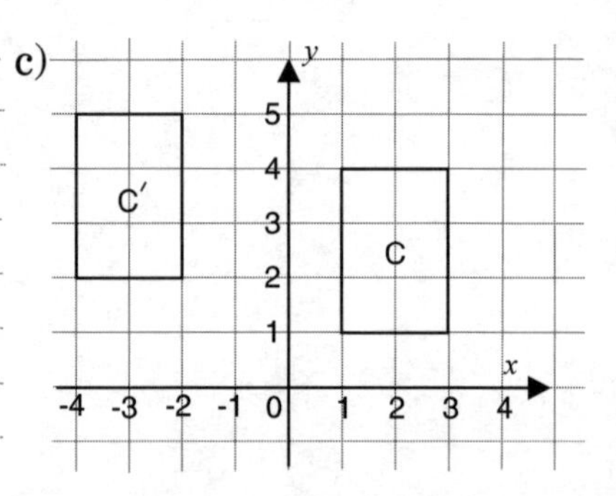

d)
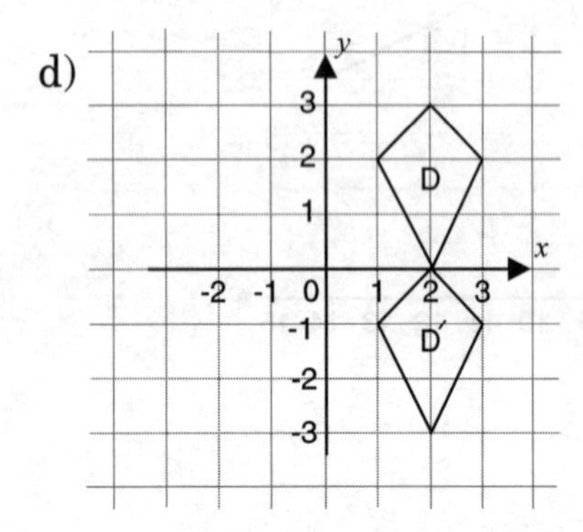

e)
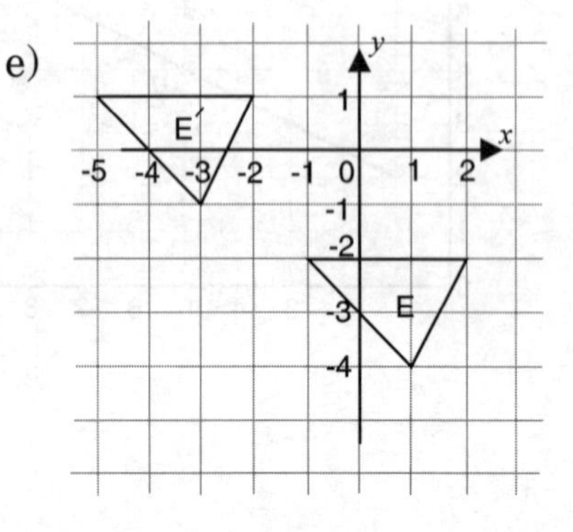

f)
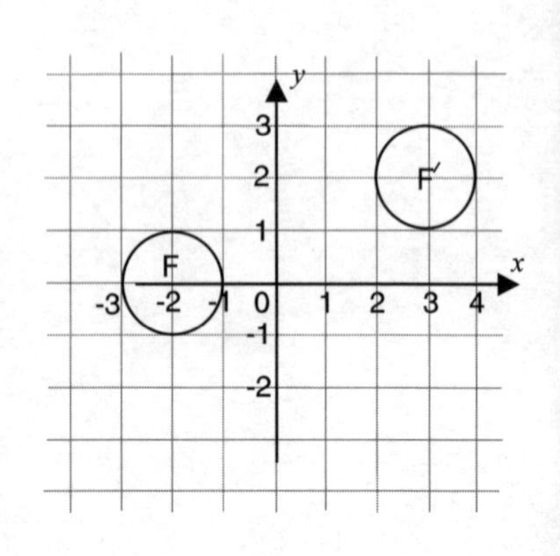

Check your answers at the end of this module.

Equal vectors

Equal vectors have the same size (magnitude) and the same direction.

As vectors are usually independent of position they can start at any point. The same vector can be at several places in a diagram.

In the figure below, **AB, CD, XY, LM** and **RS** are equal vectors.

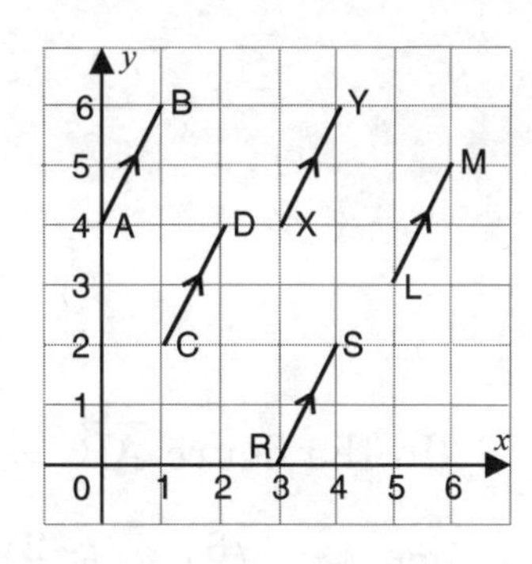

$\overrightarrow{AB} = \overrightarrow{CD} = \overrightarrow{XY} = \overrightarrow{LM} = \overrightarrow{RS} = \binom{1}{2}$.

Multiplication of a vector by a number (scalar)

A vector can be multiplied by a number. The number is sometimes called a **scalar**.

Vector $\underset{\sim}{a}$ multiplied by 2 is the vector $2\underset{\sim}{a}$. Vector $2\underset{\sim}{a}$ is twice as long as vector $\underset{\sim}{a}$. Vectors $\underset{\sim}{a}$ and $2\underset{\sim}{a}$ have the same direction. That is, they are either parallel or in the same straight line.

If $\underset{\sim}{a} = \binom{-3}{2}$, then $2\underset{\sim}{a} = 2\binom{-3}{2} = \binom{-6}{4}$.

> In general, the vector $k\underset{\sim}{a}$ (where k is a positive number), is in the same direction as $\underset{\sim}{a}$ and k times as long as $\underset{\sim}{a}$.
> When k is a negative number, the vector $k\underset{\sim}{a}$ is in the opposite direction to $\underset{\sim}{a}$ and k times as long as $\underset{\sim}{a}$.

Vector $\underset{\sim}{a}$ multiplied by -1 is the vector $-\underset{\sim}{a}$, opposite in direction to $\underset{\sim}{a}$ but with the same magnitude as $\underset{\sim}{a}$.

Example 1

If $\underset{\sim}{u} = \binom{8}{-4}$, find $\frac{1}{4}\underset{\sim}{u}$.

Solution

$$\frac{1}{4}\underset{\sim}{u} = \frac{1}{4}\binom{8}{-4} = \begin{pmatrix} \frac{1}{4} \times 8 \\ \frac{1}{4} \times (-4) \end{pmatrix} = \binom{2}{-1}$$

Example 2

If $\underset{\sim}{v} = \binom{-3}{2}$, find $-5\underset{\sim}{v}$.

Solution

$$-5\underset{\sim}{v} = -5\binom{-3}{2} = \begin{pmatrix} (-5) \times (-3) \\ (-5) \times 2 \end{pmatrix} = \binom{15}{-10}$$

Addition of vectors

In the given figure, the translation from A to B followed by the translation from B to C is the same as the translation from A to C. We write $\overrightarrow{AB} + \overrightarrow{BC} = \overrightarrow{AC}$.

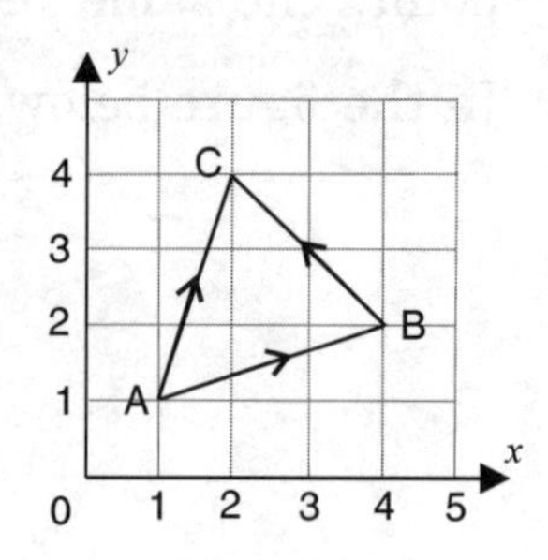

In the figure $\overrightarrow{AB} = \binom{3}{1}$, $\overrightarrow{BC} = \binom{-2}{2}$ and $\overrightarrow{AC} = \binom{1}{3}$.

That is, $\binom{3}{1} + \binom{-2}{2} = \binom{1}{3}$.

> To add vectors in number pair form, simply add the elements in corresponding positions. That is, add the top elements and then add the bottom elements.

In the figure, $\overrightarrow{AB}$ and $\overrightarrow{BC}$ are in the 'nose to tail' position, that is, the nose B of $\overrightarrow{AB}$ is connected to the tail B of $\overrightarrow{BC}$, and $\overrightarrow{AC}$ represents the sum of the vectors represented by $\overrightarrow{AB}$ and $\overrightarrow{BC}$.

Thus the rule is called the **nose to tail rule** of addition.

Since AB, BC and AC form the triangle ABC, the rule $\overrightarrow{AB} + \overrightarrow{BC} = \overrightarrow{AC}$ is also called the **triangle law**.

Example 1

In the given figure the various line segments are taken to be representatives of vectors. Find in the figure directed line segments equal to the following:

a) $\overrightarrow{AE} + \overrightarrow{EC}$
b) $\overrightarrow{DB} + \overrightarrow{BE}$
c) $\overrightarrow{AD} + \overrightarrow{DB} + \overrightarrow{BC}$
d) $\overrightarrow{CB} + \overrightarrow{BE} + \overrightarrow{EA} + \overrightarrow{AD}$

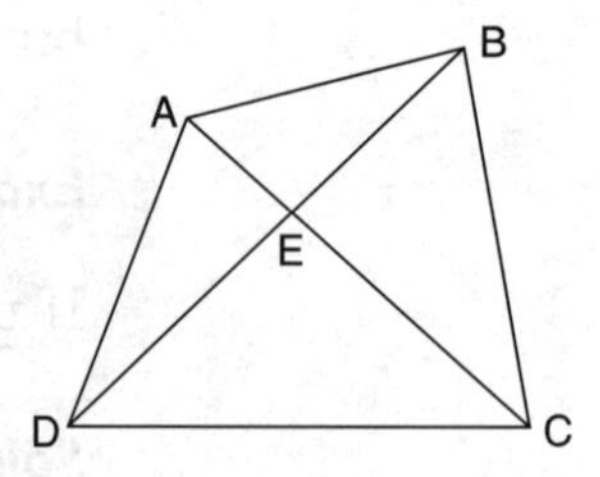

Solution

a) $\overrightarrow{AE} + \overrightarrow{EC} = \overrightarrow{AC}$
b) $\overrightarrow{DB} + \overrightarrow{BE} = \overrightarrow{DE}$
c) $\overrightarrow{AD} + \overrightarrow{DB} + \text{BC} = \overrightarrow{AB} + \text{BC} = \overrightarrow{AC}$
d) $\overrightarrow{CB} + \overrightarrow{BE} + \overrightarrow{EA} + \overrightarrow{AD} = \overrightarrow{CE} + \overrightarrow{EA} + \overrightarrow{AD} = \overrightarrow{CA} + \overrightarrow{AD} = \overrightarrow{CD}$

Example 2

If $\mathbf{a} = \binom{3}{4}$ and $\mathbf{b} = \binom{2}{-1}$, find the column vectors equal to

a + b, a − b, 3a, a + 4b, 2a − 3b.

Solution

$$\mathbf{a} + \mathbf{b} = \binom{3}{4} + \binom{2}{-1} = \binom{3\ +\ 2}{4 + (-1)} = \binom{5}{3}$$

$$\mathbf{a} - \mathbf{b} = \binom{3}{4} - \binom{2}{-1} = \binom{3\ -\ 2}{4 - (-1)} = \binom{1}{5}$$

$$\mathbf{3a} = 3 \times \binom{3}{4} = \binom{3 \times 3}{3 \times 4} = \binom{9}{12}$$

$$\mathbf{a} + \mathbf{4b} = \binom{3}{4} + 4\binom{2}{-1} = \binom{3\ +\ 8}{4 + (-4)} = \binom{11}{0}$$

$$\mathbf{2a} - \mathbf{3b} = 2\binom{3}{4} - 3\binom{2}{-1} = \binom{0}{11}$$

Example 3

OACB is a parallelogram in which $\overrightarrow{OA} = \underset{\sim}{a}$ and $\overrightarrow{OB} = \underset{\sim}{b}$.
M is the mid-point of BC, and N is the mid-point of AC.

a) Find, in terms of $\underset{\sim}{a}$ and $\underset{\sim}{b}$:

(i) $\overrightarrow{OM}$

(ii) $\overrightarrow{MN}$

b) Show that:

$\overrightarrow{OM} + \overrightarrow{MN} = \overrightarrow{OA} + \overrightarrow{AN}$

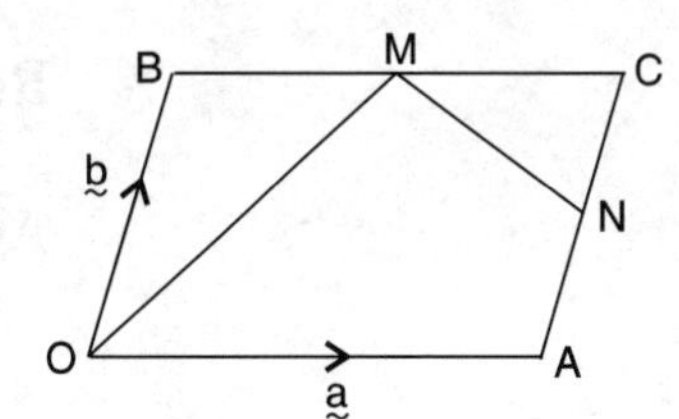

Solution

a) (i) $\overrightarrow{OM} = \overrightarrow{OB} + \overrightarrow{BM}$

$\overrightarrow{OB} = \underset{\sim}{b}$

M is the mid-point of BC, so $\overrightarrow{BM} = \frac{1}{2}\overrightarrow{BC}$.

$\overrightarrow{BM} = \frac{1}{2}\overrightarrow{BC} = \frac{1}{2}\underset{\sim}{a}$

so $\overrightarrow{OM} = \underset{\sim}{b} + \frac{1}{2}\underset{\sim}{a}$

the opposite sides of a parallelogram are equal and parallel, so $\overrightarrow{BC} = \overrightarrow{OA} = \underset{\sim}{a}$

(ii) $\overrightarrow{MN} = \overrightarrow{MC} + \overrightarrow{CN} = \frac{1}{2}\overrightarrow{BC} + \frac{1}{2}\overrightarrow{CA}$

$= \frac{1}{2}\underset{\sim}{a} + (-\frac{1}{2}\underset{\sim}{b})$ $\quad (\overrightarrow{CA} = -\overrightarrow{AC} = -\overrightarrow{OB})$

$= \frac{1}{2}\underset{\sim}{a} - \frac{1}{2}\underset{\sim}{b} = \frac{1}{2}(\underset{\sim}{a} - \underset{\sim}{b})$

b) $\overrightarrow{OM} + \overrightarrow{MN} = (\underset{\sim}{b} + \frac{1}{2}\underset{\sim}{a}) + (\frac{1}{2}\underset{\sim}{a} - \frac{1}{2}\underset{\sim}{b}) = \underset{\sim}{a} + \frac{1}{2}\underset{\sim}{b}$

$\overrightarrow{OA} + \overrightarrow{AN} = \underset{\sim}{a} + \frac{1}{2}\underset{\sim}{b}$

So $\overrightarrow{OM} + \overrightarrow{MN} = \overrightarrow{OA} + \overrightarrow{AN}$

Alternative method

Using the nose to tail rule, $\overrightarrow{OM} + \overrightarrow{MN} = \overrightarrow{ON}$

and $\overrightarrow{OA} + AN = \overrightarrow{ON}$

So $\overrightarrow{OM} + \overrightarrow{MN} = \overrightarrow{OA} + \overrightarrow{AN}$

Subtraction of vectors

You already know that $5 - 3 = 5 + (-3)$ and $x - y = x + (-y)$.

Similarly, if $\underset{\sim}{a} = \begin{pmatrix}5\\3\end{pmatrix}$ and $\underset{\sim}{b} = \begin{pmatrix}2\\1\end{pmatrix}$ then,

$\underset{\sim}{a} - \underset{\sim}{b} = \begin{pmatrix}5\\3\end{pmatrix} - \begin{pmatrix}2\\1\end{pmatrix} = \begin{pmatrix}5\\3\end{pmatrix} + \left(-\begin{pmatrix}2\\1\end{pmatrix}\right) = \begin{pmatrix}3\\2\end{pmatrix}$

So $\underset{\sim}{a} - \underset{\sim}{b} = \underset{\sim}{a} + (-\underset{\sim}{b})$

Subtracting a vector is the same as adding its negative.

Consider $\overrightarrow{AC} - \overrightarrow{AB}$:

Adding the negative of $\overrightarrow{AB}$ is the same as adding $\overrightarrow{BA}$.

$\overrightarrow{AC} - \overrightarrow{AB} = \overrightarrow{AC} + \overrightarrow{BA}$

Rearranging the vectors,

$\overrightarrow{AC} - \overrightarrow{AB} = \overrightarrow{BA} + \overrightarrow{AC}$

$= \overrightarrow{BC}$ (nose to tail rule)

The work is beginning to look quite complicated. It's not as difficult as it looks! Try these questions for yourself.

EXERCISE 15

1. $\mathbf{p} = \begin{pmatrix}4\\-2\end{pmatrix}$ and $\mathbf{q} = \begin{pmatrix}-1\\-3\end{pmatrix}$.

   Express in column vector form:

   a) $3\mathbf{p}$ b) $\mathbf{p} + \mathbf{q}$

2. Given that $\underset{\sim}{a} = \begin{pmatrix}3\\-2\end{pmatrix}$ and $\underset{\sim}{b} = \begin{pmatrix}-4\\3\end{pmatrix}$, express $2\underset{\sim}{a} - \underset{\sim}{b}$ as a column vector.

3. In the diagram BCE and ACD are straight lines. $\overrightarrow{AB} = 2\mathbf{a}$ and $\overrightarrow{BC} = 3\mathbf{b}$. The point C divides AD in the ratio 2 : 1 and divides BE in the ratio 3 : 1. Express, in terms of **a** and **b**, the vectors:

   a) $\overrightarrow{AC}$
   b) $\overrightarrow{CD}$
   c) $\overrightarrow{CE}$
   d) $\overrightarrow{ED}$

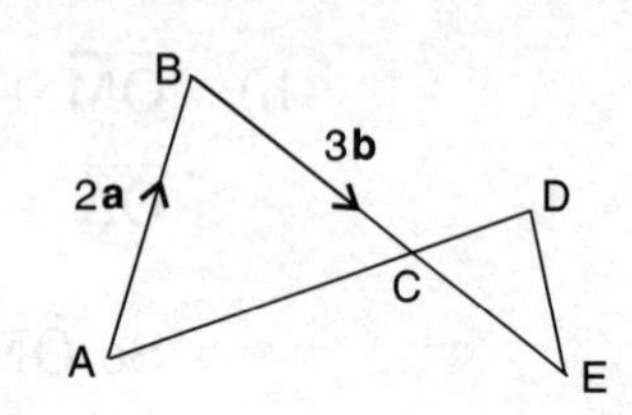

4. In the diagram, OC is parallel to AB and OC = 2AB. Given that $\overrightarrow{OA} = \underset{\sim}{a}$ and $\overrightarrow{OB} = \underset{\sim}{b}$, find, in terms of $\underset{\sim}{a}$ and $\underset{\sim}{b}$:
   a) $\overrightarrow{AB}$
   b) $\overrightarrow{OC}$
   c) $\overrightarrow{BC}$

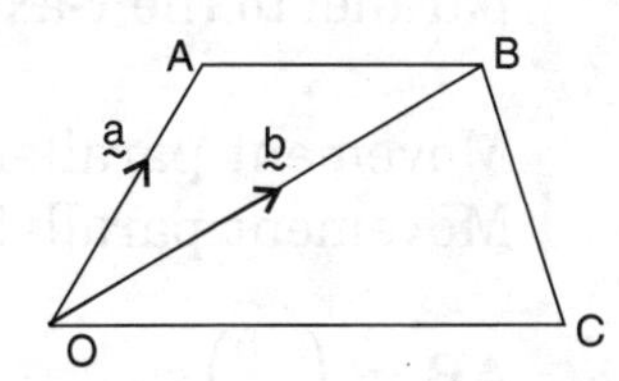

Check your answers at the end of this module.

If you are following the CORE syllabus, you should now move on to the 'Summary'. The remainder of Unit 3 covers items in the EXTENDED syllabus.

Position vectors

A vector can start from any point because any line segment with the correct magnitude and direction can represent it. A vector starting from the origin, O, is called a **position vector**.

In the given figure the point A has position vector $\underset{\sim}{a}$.

If $\underset{\sim}{a} = \begin{pmatrix}3\\2\end{pmatrix}$, then the coordinates of A will be (3, 2).

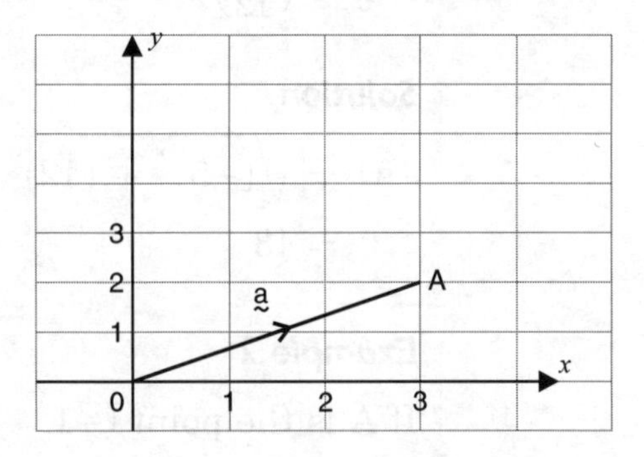

Example

The position vector of A is $\begin{pmatrix}3\\2\end{pmatrix}$ and the position vector of B is $\begin{pmatrix}-2\\4\end{pmatrix}$. Find the vector 2 **AB**.

Solution

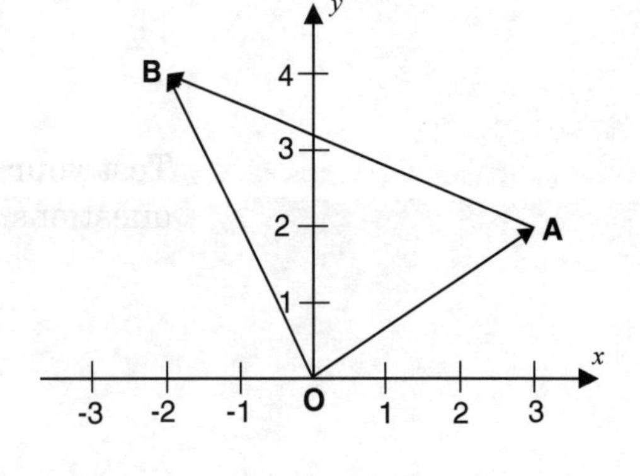

$\mathbf{OA} = \begin{pmatrix}3\\2\end{pmatrix}$ $\mathbf{OB} = \begin{pmatrix}-2\\4\end{pmatrix}$

$\mathbf{AB} = -\ \mathbf{OA} + \mathbf{OB}$

$= \mathbf{OB} - \mathbf{OA}$

That is, $\mathbf{AB} = \begin{pmatrix}-2\\4\end{pmatrix} - \begin{pmatrix}3\\2\end{pmatrix} = \begin{pmatrix}-5\\2\end{pmatrix}$

So $2\ \mathbf{AB} = 2\begin{pmatrix}-5\\2\end{pmatrix} = \begin{pmatrix}-10\\4\end{pmatrix}$

We could also have found the column vector for **AB**, the displacement (translation) from A to B, by counting the movement parallel to the x-axis followed by the movement parallel to the y-axis.

Movement parallel to the x-axis = –5 units.
Movement parallel to the y-axis = 2 units.

$\mathbf{AB} = \begin{pmatrix} -5 \\ 2 \end{pmatrix}$

Thus $2\,\mathbf{AB} = \begin{pmatrix} -10 \\ 4 \end{pmatrix}$

The magnitude of a vector

The length of a vector is normally called the **magnitude** or **modulus** of the vector.

The magnitude of vector **AB** is written as $|\mathbf{AB}|$.

The magnitude of $\overrightarrow{AB}$ is $|\overrightarrow{AB}|$ and the magnitude of $\underset{\sim}{a}$ is $|\underset{\sim}{a}|$.

If $\overrightarrow{AB} = \begin{pmatrix} x \\ y \end{pmatrix}$, then $|AB| = \sqrt{(x^2 + y^2)}$.

make sure you can see how Pythagoras's theorem is used to get this result

Example 1

If $\underset{\sim}{a} = \begin{pmatrix} -5 \\ 12 \end{pmatrix}$, find $|\underset{\sim}{a}|$.

Solution

$$|\underset{\sim}{a}| = \sqrt{(-5)^2 + (12)^2} = \sqrt{169}$$
$$= 13$$

Example 2

If A is the point (–1, –2) and B is (5, 6), find $|\overrightarrow{AB}|$.

Solution

$$\overrightarrow{OA} = \begin{pmatrix} -1 \\ -2 \end{pmatrix} \quad \overrightarrow{OB} = \begin{pmatrix} 5 \\ 6 \end{pmatrix}$$

$$\overrightarrow{AB} = \overrightarrow{AO} + \overrightarrow{OB} = -\overrightarrow{OA} + \overrightarrow{OB} = \begin{pmatrix} -(-1)+5 \\ -(-2)+6 \end{pmatrix} = \begin{pmatrix} 6 \\ 8 \end{pmatrix}$$

$$|\overrightarrow{AB}| = \sqrt{6^2 + 8^2} = 10$$

Test your understanding of this work by answering the following questions.

EXERCISE 16

1. O is the point (0, 0), P is (3, 4), Q is (−5, 12) and R is (−8, −15). Find the values of $|\overrightarrow{OP}|$, $|\overrightarrow{OQ}|$ and $|\overrightarrow{OR}|$.

2. $\overrightarrow{OA} = \begin{pmatrix} 4 \\ 2 \end{pmatrix}$, $\overrightarrow{OB} = \begin{pmatrix} -1 \\ 3 \end{pmatrix}$ and $\overrightarrow{OC} = \begin{pmatrix} 6 \\ -2 \end{pmatrix}$.

   a) Write down the coordinates of A, B and C.
   b) Write down the vectors $\overrightarrow{AB}$, $\overrightarrow{CB}$ and $\overrightarrow{AC}$ in column vector form.

3. OACB is a parallelogram in which $\overrightarrow{OA} = 2\underset{\sim}{p}$ and $\overrightarrow{OB} = 2\underset{\sim}{q}$. M is the mid-point of BC, and N is the mid-point of AC.

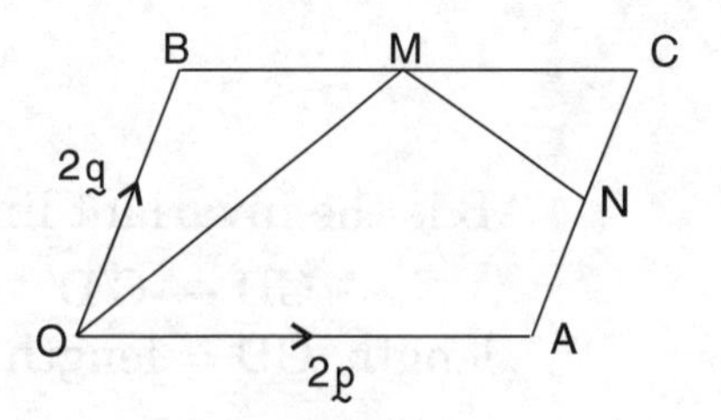

   Find, in terms of $\underset{\sim}{p}$ and $\underset{\sim}{q}$:
   a) $\overrightarrow{AB}$
   b) $\overrightarrow{ON}$
   c) $\overrightarrow{NM}$

4. OATB is a parallelogram.
   M, N, P are the mid-points of BT, AT and MN respectively.
   O is the origin, and the position vectors of A and B are $\underset{\sim}{a}$ and $\underset{\sim}{b}$ respectively.

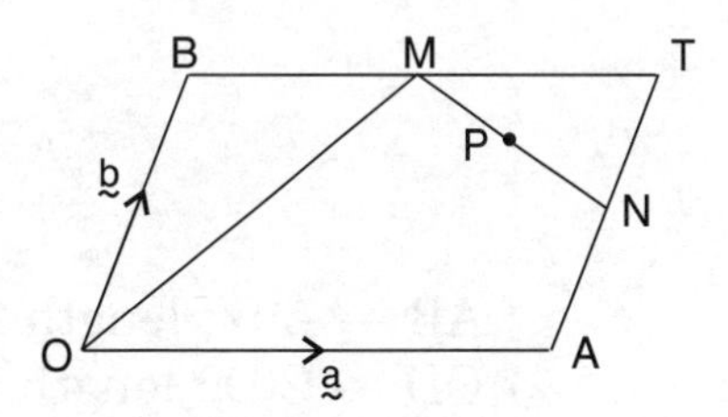

   Find, in terms of $\underset{\sim}{a}$ and/or $\underset{\sim}{b}$:
   a) $\overrightarrow{MT}$
   b) $\overrightarrow{TN}$
   c) $\overrightarrow{MN}$
   d) the position vector of P, giving your answer in its simplest form.

Check your answers at the end of this module.

C Further transformations

Shear

The transformation that keeps one line fixed, moves all other points parallel to this line and maps every straight line onto a straight line is called a **shear**. The fixed line is called the **invariant line**.
The invariant line can be anywhere, in the figure or outside the figure. A shear is not an isometry, because the shape of the figure is distorted.

In Figure 1, the rectangle ABCD maps onto ABC′D′ under a shear.
ABCD → ABC′D′

AB → AB

Figure 1

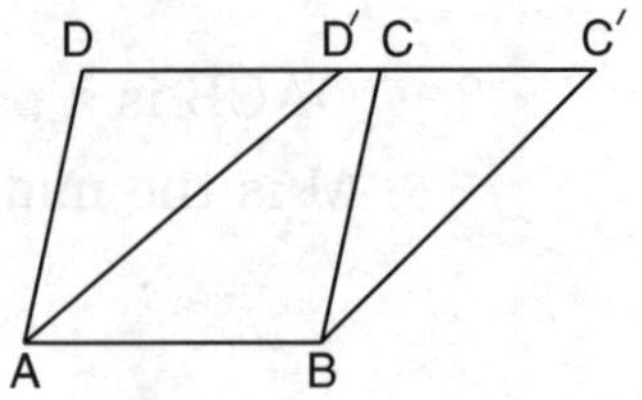

AB is the invariant line. That is, the points on the line AB are fixed.

CD → C′D′

length CD = length C′D′

All the points (except those on AB) move parallel to the invariant line AB.

Area of ABCD = area of ABC′D′.

In Figure 2, the invariant line is PQ.
ABCD → A′B′C′D′

Figure 2

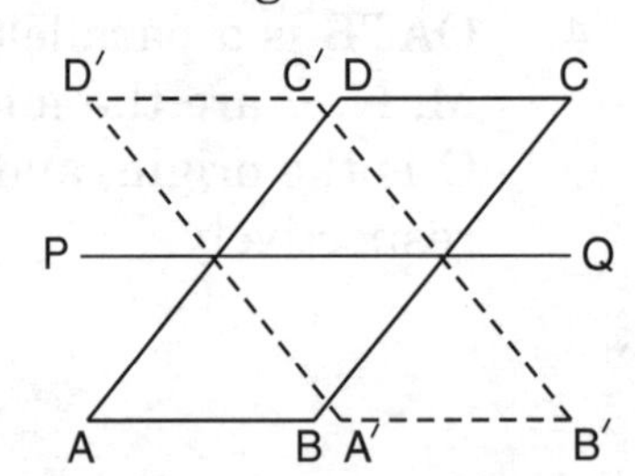

AB → A′B′; length AB = length A′B′
CD → C′D′; length CD = length C′D′

The points on one side of the invariant line move to the left while the points on the other side move to the right.

Area of ABCD = area of A′B′C′D′ .

Properties of shear

- The points on the invariant line are fixed. In the diagram below, PQ is the invariant line. P, Q and all points on the line PQ are unaffected by the shear.
- All the points that are not on the invariant line move parallel to the invariant line.

  C moves to C′ in the direction parallel to PQ.

  X moves to X′ in the direction parallel to PQ.

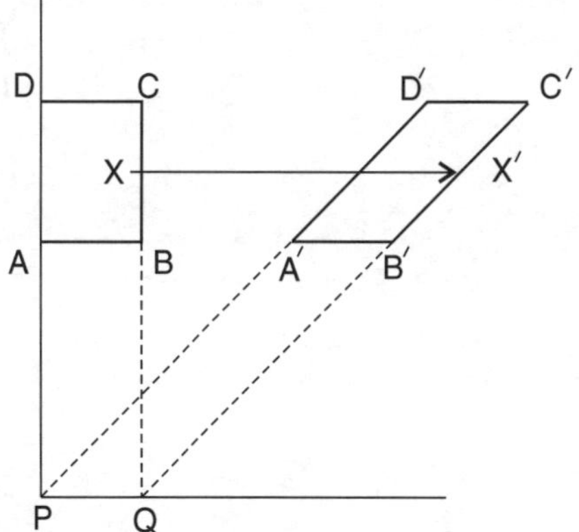

- The perpendicular distance of points from the invariant line remains constant.

  Perpendicular distance of D from PQ = perpendicular distance of D′ from PQ.

  to find the perpendicular distance of D′ from PQ you need to extend PQ

  Perpendicular distance of X from PQ = perpendicular distance of X′ from PQ.

- The image of a straight line is a straight line.
- A straight line and its image meet at a point on the invariant line, unless the line is parallel to the invariant line. See how DA and D′A′ produced meet at P.
- Points on the opposite sides of the invariant line are displaced in opposite directions.
- The area of a figure remains constant.
- The distance moved by a point is proportional to its distance from the invariant line. In other words, the ratio of the distance moved by a point to its distance from the invariant line is constant.
  This constant is called the **shear factor**.
  PQ is the invariant line.
  A maps onto A′ under a shear.

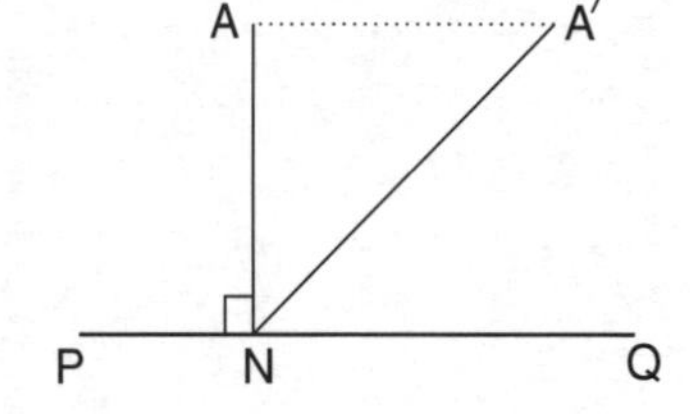

$$\text{Shear factor} = \frac{\text{distance moved by a point}}{\text{distance of the point from the invariant line}} = \frac{AA'}{AN}$$

> To define a shear we need to know only the invariant line and the image of a point not on the invariant line.

Example 1

In the figure, the rectangle OABC is mapped onto parallelogram OAB′C′ under a shear.
Find the invariant line and the shear factor.

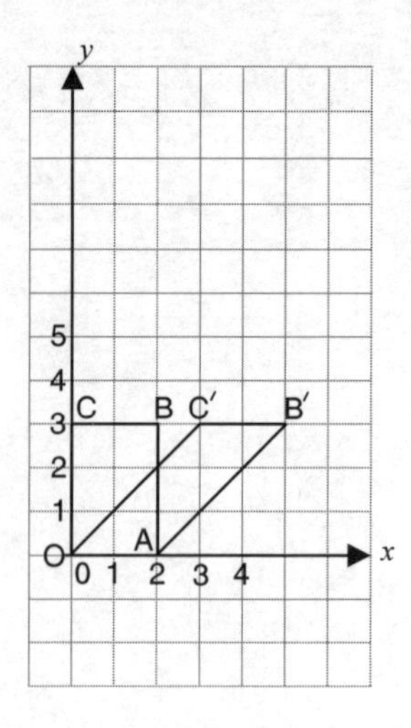

Solution

The points on OA do not move, so OA is the invariant line.
C has moved onto C′. The distance of C from OA $= 3$ units. The distance moved by C $= 3$ units. The ratio,
the distance moved by C : the distance of C from OA $= 3 : 3 = 1$.
Therefore the shear factor $= 1$.

Example 2

Find the image of ABCD under a shear which keeps the x-axis fixed (invariant) and maps A to A′.

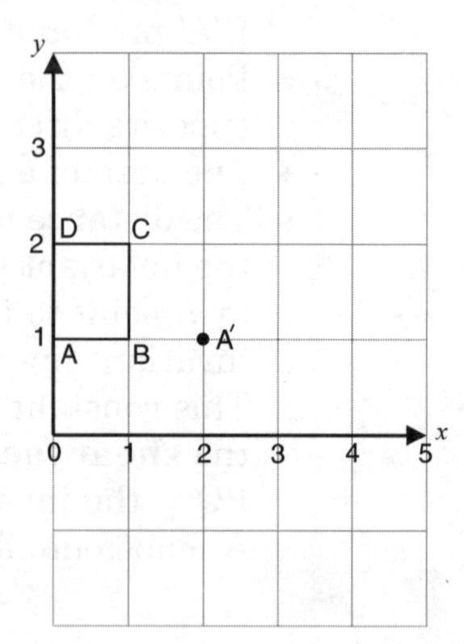

Solution

A moves 2 units to A′.
Distance of A from the invariant line (the x-axis) $= 1$ unit.

$$\frac{\text{distance moved by A}}{\text{distance of A from } x\text{–axis (invariant line)}} = \frac{2}{1} = 2, \text{ so the shear factor} = 2$$

$$\frac{\text{distance moved by B}}{\text{distance of B from } x\text{–axis (invariant line)}} = \text{shear factor} = 2$$

Distance moved by B $= 2 \times$ distance of B from the x-axis.
Distance of B from the x-axis $= 1$ unit.
Distance moved by B $= 2 \times 1 = 2$ units.
Coordinates of B′ (the image of B) $= (3, 1)$.

$$\frac{\text{distance moved by C}}{\text{distance of C from } x\text{–axis (invariant line)}} = \text{shear factor} = 2$$

Distance moved by C $= 2 \times$ distance of C from the x-axis.
Distance of C from x-axis $= 2$ units.
Distance moved by C $= 2 \times 2 = 4$ units.
Coordinates of C′ (the image of C) $= (5, 2)$.

$$\frac{\text{distance moved by D}}{\text{distance of D from } x\text{–axis (invariant line)}} = \text{shear factor} = 2$$

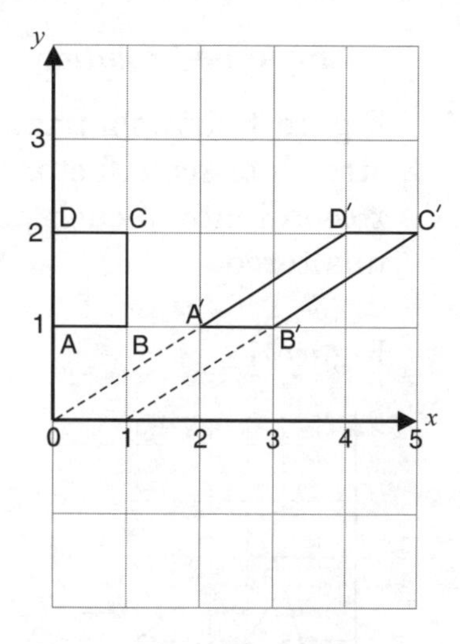

Distance of D from x-axis $= 2$ units.

Distance moved by D $= 4$ units.

Coordinates of D′ (the image of D) $= (4, 2)$.

The image of the square ABCD is the parallelogram A′B′C′D′ shown in the diagram.

Stretch

A **one-way stretch** is an enlargement in one direction from a given line. The given line is called the invariant line or the axis.

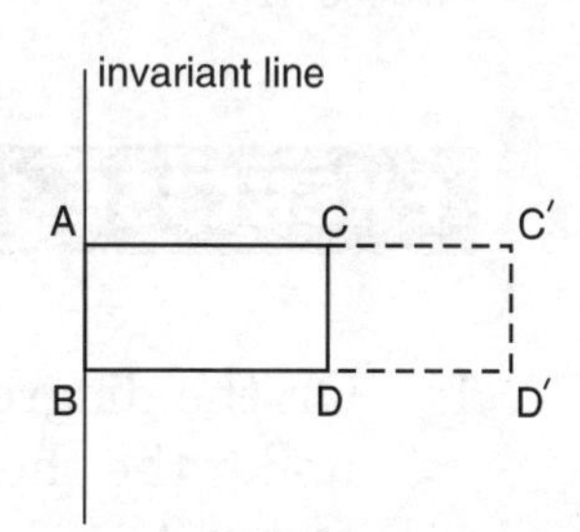

$$\text{The scale factor} = \frac{\text{distance of the image of a point from the invariant line}}{\text{distance of the point from the invariant line}}$$

$$\text{Scale factor} = \frac{AC'}{AC} = \frac{BD'}{BD}$$

For a one-way stretch with the y-axis as invariant line, the y-coordinates remain unaltered and the x-coordinates are multiplied by the scale factor.

In a one-way stretch with the x-axis as invariant line, the x-coordinates remain unaltered and the y-coordinates are multiplied by the scale factor.

Properties of a stretch

- A one-way stretch with scale factor (−1) is equivalent to a reflection in the invariant line.
- A **two-way stretch** is a combination of two one-way stretches with perpendicular invariant lines. The only invariant point is the point of intersection of the two lines. A two-way stretch affects both coordinates of each point by the respective scale factors.
- A two-way stretch with equal scale factors is an enlargement, with centre of enlargement at the point of intersection of the invariant lines, and the same scale factor.

Example and solution

Figure 1 shows a triangle ABC and its image A′B′C′ after a one-way stretch of scale factor 2 with the y-axis as invariant line. The x-coordinate of each point is doubled while the y-coordinate is unaltered.

Figure 1

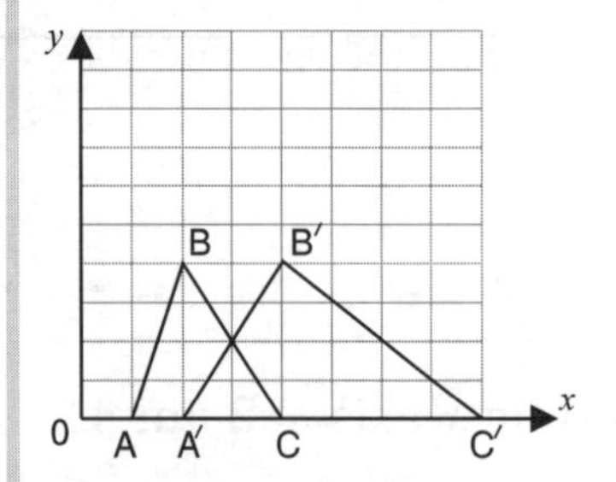

Figure 2

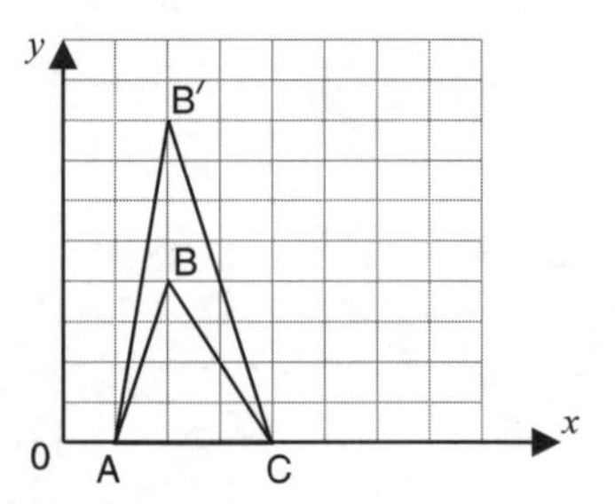

Figure 2 shows ABC and its image AB′C after a one-way stretch of scale factor 2 with the x-axis as invariant line.

The y-coordinate of each point is multiplied by 2 while the x-coordinate is unaltered.

Here are three questions on shear and stretch transformations for you to try.

EXERCISE 17

1. On the diagram, draw the image A′B′C′ of the triangle ABC under the shear which has the x-axis as invariant line and shear factor 2.

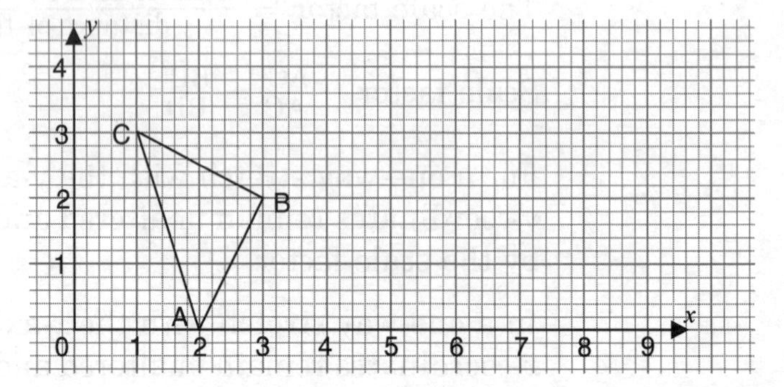

2. On the diagram, draw the image P′Q′R′ of the triangle PQR under the shear which has the line l as invariant line and maps P onto the point P′ shown.

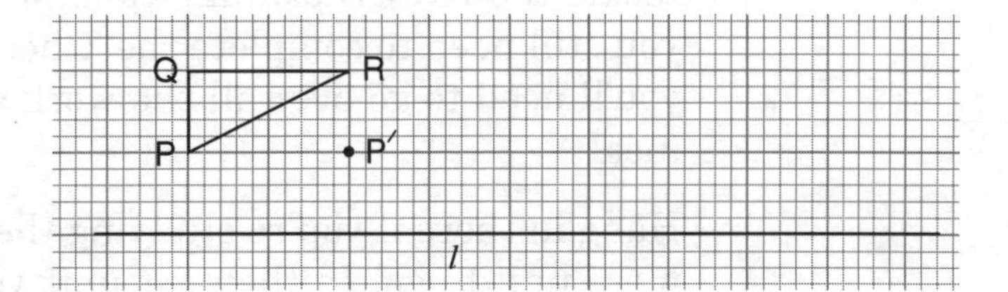

3. a) Describe the single transformation which maps the rectangle OABC onto parallelogram OAPQ.
   b) Describe the single transformation which maps the rectangle OABC onto the rectangle OAKL.

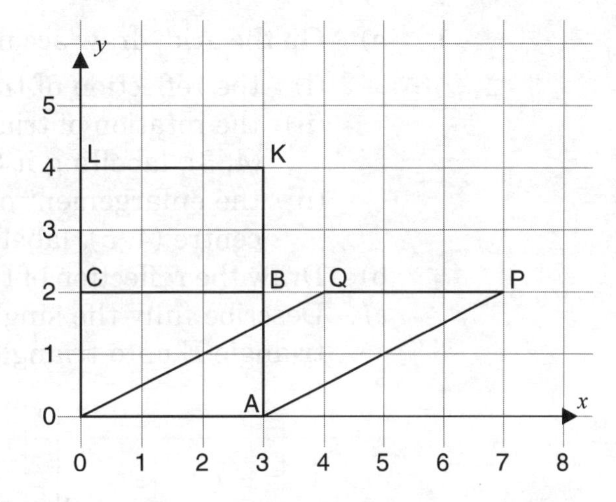

Check your answers at the end of this module.

Summary

You've learnt about a number of transformations in this unit and you'll probably have to go over this work quite a few times before you are able to remember the different types and their properties.

You should be familiar with these three isometric transformations:

- translation
- reflection
- rotation

You also learnt about the non-isometric transformation:

- enlargement

In Section B I introduced you to the topic of vectors and besides being able to work with vectors I showed you how vectors can be used to describe translation.

Those studying the EXTENDED syllabus should know what a position vector is and how to calculate the magnitude of a vector as well. The two transformations you should also know about are:

- shear
- stretch

Well done, if you are studying the CORE syllabus. You have now completed the final unit of the final module of this course. I hope that you have learnt many new ideas and that you feel well prepared to tackle the IGCSE examination. Maybe not yet because, of course, you still need to complete the 'Check your progress' for this unit and you'll need to go over all the work a few more times, I'm sure. Good luck!

Hang in there if you're studying the EXTENDED syllabus – just one more unit to go! In the next unit you'll be tackling the topic of matrices.

Check your progress

1. a) On the grid, draw accurately the following transformations:
   (i) the reflection of triangle A in the y-axis, labelling it B
   (ii) the rotation of triangle A through 180° about the point (4, 3), labelling it C
   (iii) the enlargement of triangle A, scale factor 2, centre (4, 5), labelling it D

   b) Draw the reflection of triangle B in the x-axis, labelling it E.

   c) Describe fully the single transformation which maps triangle E onto triangle C.

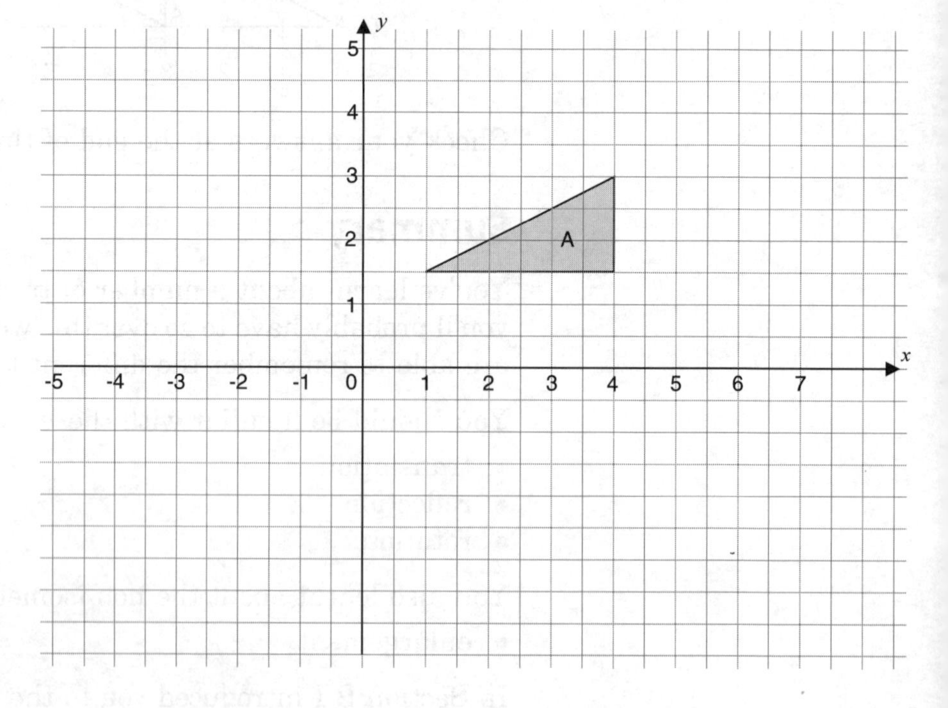

2. $\underset{\sim}{m} = \begin{pmatrix} 3 \\ -4 \end{pmatrix}$ and $\underset{\sim}{n} = \begin{pmatrix} -2 \\ 1 \end{pmatrix}$.

a) Find:
   (i) $\underset{\sim}{m} + \underset{\sim}{n}$
   (ii) $3\underset{\sim}{n}$

b) Draw the vector $\underset{\sim}{m}$ on the grid.

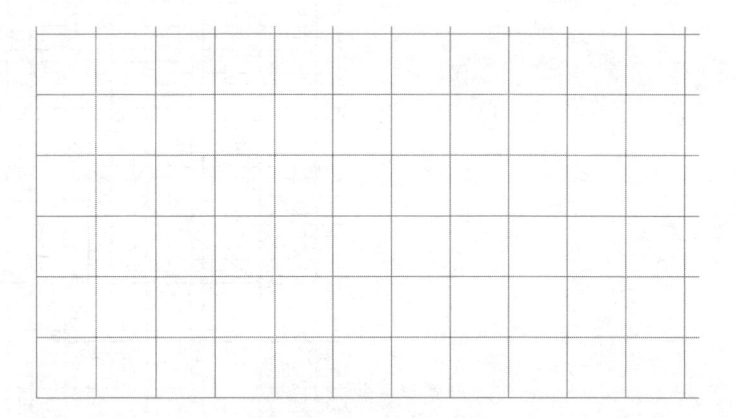

3. Describe fully the transformations of the shaded triangle on to triangles A, B, C and D.

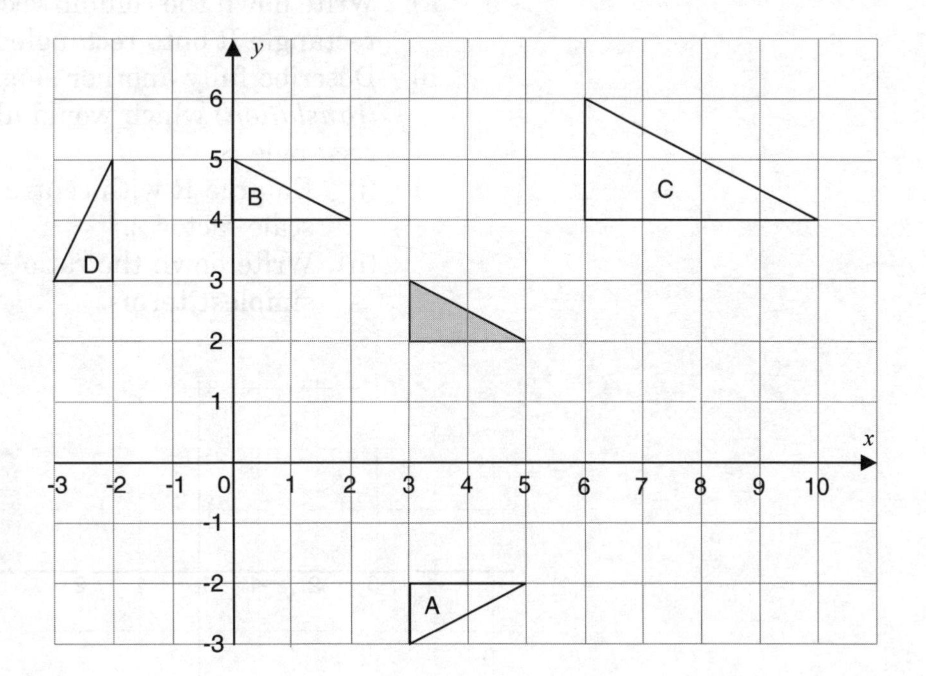

4. Triangle ABC is mapped onto triangle A′B′C′ by an enlargement.
   a) Find the centre of the enlargement.
   b) What is the scale factor of the enlargement?

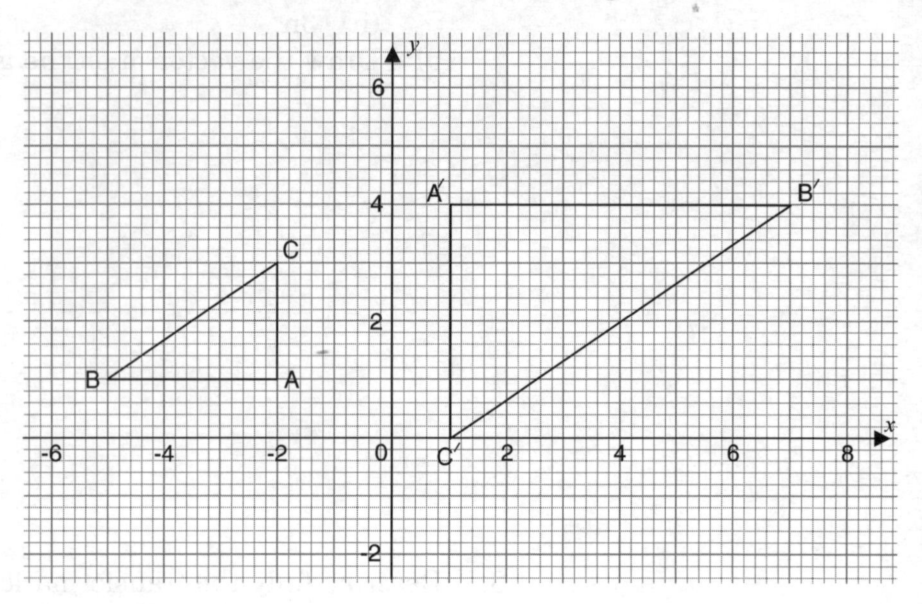

5. a) Write down the column vector of the translation that maps rectangle R onto rectangle S.
   b) Describe fully another single transformation (not a *translation*) which would also map rectangle R onto rectangle S.
   c) (i) Enlarge R with centre of enlargement A (10, 2) and scale factor 2.
   (ii) Write down the ratio $\frac{\text{area of enlarged rectangle}}{\text{area of rectangle R}}$ in its simplest terms.

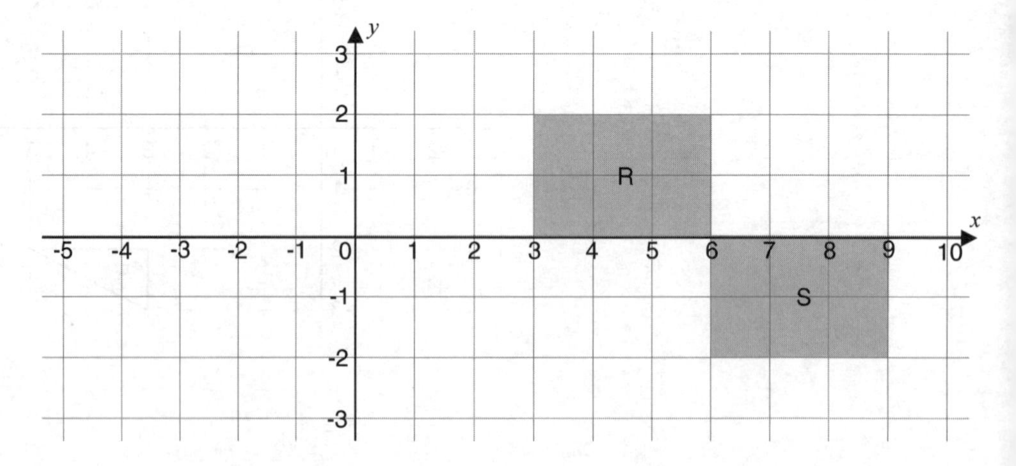

6. $\overrightarrow{OA} = \underset{\sim}{a}$ and $\overrightarrow{OB} = \underset{\sim}{b}$.

a) $\overrightarrow{OC} = \underset{\sim}{a} + 2\underset{\sim}{b}$. Label the point C on the diagram.

b) D = (0, −1). Write $\overrightarrow{OD}$ in terms of $\underset{\sim}{a}$ and $\underset{\sim}{b}$.

c) Calculate $|\underset{\sim}{a}|$, giving your answer to 2 decimal places.

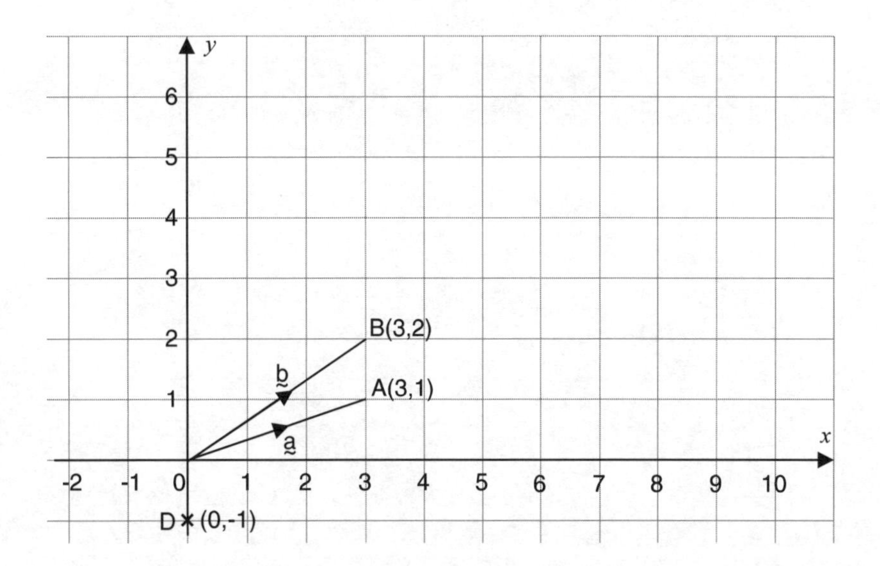

7. a) In each case, describe fully the single transformation which maps A onto:

(i) B (ii) C (iii) D (iv) E (v) F

b) State which shapes have an area equal to that of A.

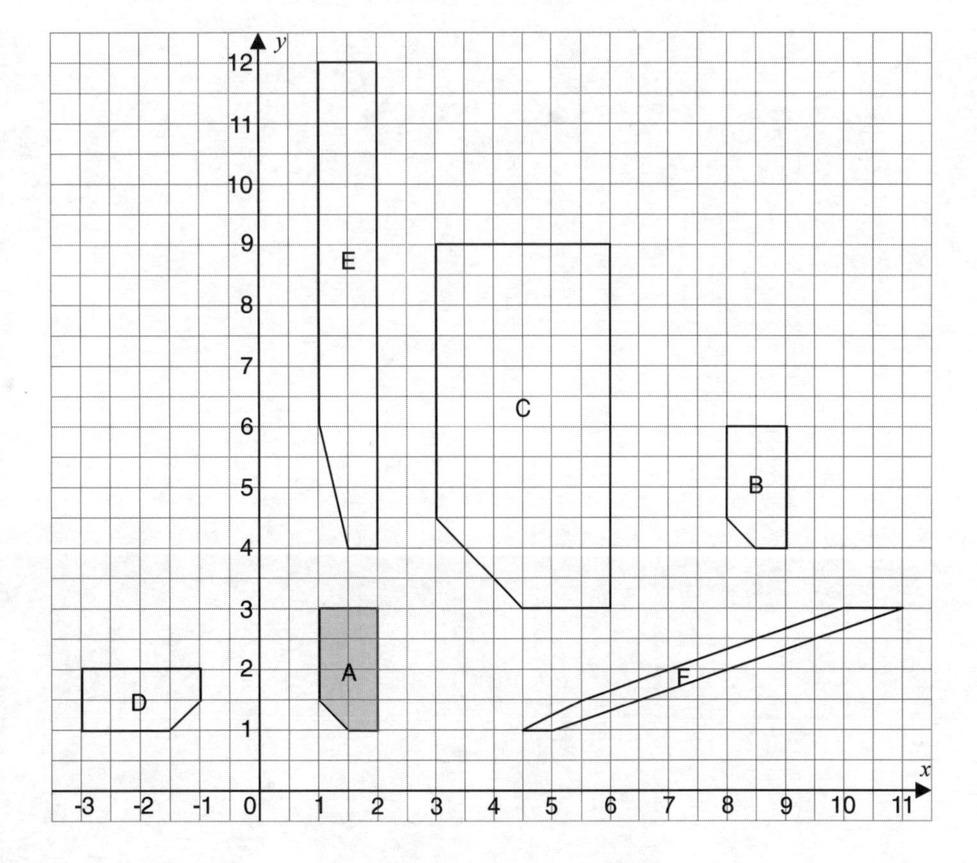

Check your answers at the end of this module.

Unit 4
Matrices and Matrix Transformations

In this unit I will begin by explaining to you what a matrix is and how to work with matrices. Once you are familiar with matrices you'll learn how they can be used to describe geometrical transformations.

This unit is divided into three sections:

Section	Title	Time
A	Matrices to hold information	$\frac{1}{2}$ hour
B	Describing and working with matrices	6 hours
C	Using matrices for geometrical transformations	4 hours

By the end of this unit, you should be able to:

- display information in the form of a matrix
- define the order of a matrix
- know the different types of matrix
- add and subtract matrices
- multiply a matrix by a scalar
- multiply matrices
- use matrices to represent geometrical transformations
- find the matrices of common geometrical transformations
- carry out successive geometrical transformations in the correct order.

A Matrices to hold information

Mary is a schoolgirl. Every week she buys the following items. On Mondays she buys 2 loaves of bread, 10 eggs, 3 litres of milk and 1 jar of jam. Again, she has to buy 2 loaves of bread, 8 eggs and 2 litres of milk on Wednesdays. This routine repeats on Fridays with 4 loaves of bread, 12 eggs, 3 litres of milk and 1 jar of jam. This is her weekly routine.

How can she write this information in a better way so that she can get the details at a glance? Oh yes, she can put all these pieces of information in a table as shown on the next page.

	Monday	Wednesday	Friday
Bread	2	2	4
Egg	10	8	12
Milk	3	2	3
Jam	1	0	1

Can you suggest how we can simplify this further?

Since this is her weekly routine, we can leave out the days and the names of the items.

The information could be shown as:

$$\begin{array}{ccc} 2 & 2 & 4 \\ 10 & 8 & 12 \\ 3 & 2 & 3 \\ 1 & 0 & 1 \end{array}$$

You can put this information either within circular brackets or rectangular brackets.

$$\begin{pmatrix} 2 & 2 & 4 \\ 10 & 8 & 12 \\ 3 & 2 & 3 \\ 1 & 0 & 1 \end{pmatrix} \text{ or } \begin{bmatrix} 2 & 2 & 4 \\ 10 & 8 & 12 \\ 3 & 2 & 3 \\ 1 & 0 & 1 \end{bmatrix}$$

An array of numbers arranged in the above form is called a **matrix.** (The plural of matrix is matrices.)

We can also use matrices to show information about the number of routes between towns.

The figure below shows the towns P, Q and R and the bus routes.

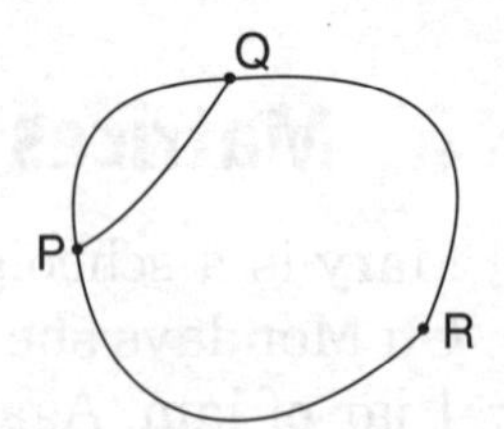

The connecting lines show the routes and they can be of any shape.

From the figure we can gather information such as:

There are two direct routes from P to Q.

There is one direct route from P to R.

There is one direct route from Q to R.

We can represent this information in a table as shown below.

		TO		
		P	Q	R
	P	0	2	1
FROM	Q	2	0	1
	R	1	1	0

If we omit the words 'FROM' and 'TO' and the letters P, Q, R, we can present the information in a matrix like this:

$$\begin{pmatrix} 0 & 2 & 1 \\ 2 & 0 & 1 \\ 1 & 1 & 0 \end{pmatrix}$$

B Describing and working with matrices

A set of numbers arranged in rows and columns enclosed in round or square brackets is called a **matrix**. I will use round brackets.

The matrix $\begin{pmatrix} 2 & 2 & 4 \\ 10 & 8 & 12 \\ 3 & 2 & 3 \\ 1 & 0 & 1 \end{pmatrix}$ has 4 rows and 3 columns.

Column 1 ↓ Column 2 ↓ Column 3 ↓

$$\begin{matrix} \text{Row 1} \rightarrow \\ \text{Row 2} \rightarrow \\ \text{Row 3} \rightarrow \\ \text{Row 4} \rightarrow \end{matrix} \begin{pmatrix} 2 & 2 & 4 \\ 10 & 8 & 12 \\ 3 & 2 & 3 \\ 1 & 0 & 1 \end{pmatrix}$$

Each number in the array is called an **entry** or an **element** of the matrix.

A matrix is often denoted by a capital letter.
For example, we can call the above matrix A.

$$A = \begin{pmatrix} 2 & 2 & 4 \\ 10 & 8 & 12 \\ 3 & 2 & 3 \\ 1 & 0 & 1 \end{pmatrix}$$

The order of a matrix

The **order of a matrix** gives the number of rows followed by the number of columns in a matrix. A matrix with 3 rows and 2 columns is a 3 by 2 matrix or a 3×2 matrix. The order of such a matrix is 3×2.

A matrix of order 4×5 has 4 rows and 5 columns.

The matrix $\begin{pmatrix} 2 & 0 & 5 \\ 1 & 3 & 4 \end{pmatrix}$ is of the order 2×3.

you say '2 by 3'

Types of matrix

Square matrix

A matrix with equal numbers of rows and columns is called a **square matrix**.

$$P = \begin{pmatrix} 3 & 1 \\ 2 & 4 \end{pmatrix}$$

The matrix P has 2 rows and 2 columns. Therefore, it is a square matrix of order 2×2.

$$Q = \begin{pmatrix} 4 & 1 & 0 \\ 5 & 3 & 1 \\ 1 & 2 & 5 \end{pmatrix}$$

The matrix Q is a square matrix of order 3×3.

Diagonal matrix

A square matrix that has all its elements zero, except those on the leading diagonal (top left to the bottom right), is called a **diagonal matrix**.

$$C = \begin{pmatrix} 4 & 0 & 0 \\ 0 & 3 & 0 \\ 0 & 0 & 5 \end{pmatrix}$$

The leading diagonal elements are 4, 3 and 5. The other elements are zeros.

Unit matrix

A diagonal matrix with its leading diagonal elements equal to 1 is called a **unit matrix.** It is denoted by the letter I.

$$I = \begin{pmatrix} 1 & 0 \\ 0 & 1 \end{pmatrix}$$

I is a 2 by 2 unit matrix.

Zero matrix

A matrix with all its elements zero is called a **zero matrix** or a **null matrix**. It is denoted by the letter O.

$$O = \begin{pmatrix} 0 & 0 \\ 0 & 0 \end{pmatrix}$$

Equal matrices

Two matrices are said to be equal if and only if they are identical. That is, they must have the same number of rows, and the same number of columns, and the elements must respectively be equal. If these conditions are not satisfied, the matrices are not equal.

Example 1 and solution

$$A = \begin{pmatrix} 4 & 1 \\ 3 & 5 \end{pmatrix} \quad B = \begin{pmatrix} 4 & 1 \\ 3 & 5 \end{pmatrix}$$

The order of matrix A is 2×2 and that of B is 2×2.
Both have the same elements in the corresponding positions.
Therefore, they are equal.

Example 2 and solution

$$C = \begin{pmatrix} 5 & 3 \\ 2 & 4 \end{pmatrix} \quad D = \begin{pmatrix} 4 & 2 \\ 3 & 5 \end{pmatrix}$$

Both matrices are of the order 2×2.
The corresponding elements are not equal. Therefore, C is not equal to D. (The fact that both matrices have elements 2, 3, 4, 5 is irrelevant.)

EXERCISE 18

1. Write down the order of each of the following matrices.

a) $\begin{pmatrix} 6 & 3 \\ 1 & 4 \end{pmatrix}$ b) $\begin{pmatrix} 6 & 3 & 0 \\ 1 & 4 & 1 \end{pmatrix}$

c) $\begin{pmatrix} 6 & 3 \\ 1 & 4 \\ 2 & 5 \end{pmatrix}$ d) $\begin{pmatrix} 2 & 7 \end{pmatrix}$

e) $\begin{pmatrix} 1 & 3 & 4 \end{pmatrix}$ f) $\begin{pmatrix} 2 \end{pmatrix}$

g) $\begin{pmatrix} 5 \\ 1 \end{pmatrix}$ h) $\begin{pmatrix} 2 \\ 1 \\ 4 \end{pmatrix}$

i) $\begin{pmatrix} 2 & 1 & -3 \\ 3 & 4 & 0 \\ 5 & 1 & 1 \end{pmatrix}$

2. Write down the equal matrices in the following list.

$A = \begin{pmatrix} 4 & 5 \end{pmatrix}$ $B = \begin{pmatrix} 5 & 4 \end{pmatrix}$

$C = \begin{pmatrix} 4 & 5 & 6 \end{pmatrix}$ $D = \begin{pmatrix} 4 \\ 5 \\ 6 \end{pmatrix}$

$E = \begin{pmatrix} 4 & 5 & 6 \end{pmatrix}$ $G = \begin{pmatrix} 6 & 5 & 4 \end{pmatrix}$

$H = \begin{pmatrix} 6 & 3 \\ 1 & 4 \end{pmatrix}$ $J = \begin{pmatrix} 6 & 3 \\ 1 & 4 \end{pmatrix}$

$K = \begin{pmatrix} 6 & 3 \\ 1 & 4 \\ 2 & 0 \end{pmatrix}$

Check your answers at the end of this module.

Addition of matrices

Subject	Term 1	Term 2	Term 3	Year Mark
Maths	50	70	60	?
Science	60	50	80	?
English	80	40	50	?

The above table shows the marks scored by a student. It shows the number of terms, subjects and the marks. How can the student get his/her year marks? I am sure you know the answer. It is simple addition!

We get the results by setting the above information in matrix form and adding the corresponding elements.

$$\begin{pmatrix} 50 \\ 60 \\ 80 \end{pmatrix} + \begin{pmatrix} 70 \\ 50 \\ 40 \end{pmatrix} + \begin{pmatrix} 60 \\ 80 \\ 50 \end{pmatrix} = \begin{pmatrix} 180 \\ 190 \\ 170 \end{pmatrix}$$

We can generalise this principle.

> To find the sum of two or more matrices add the corresponding elements in each matrix provided the matrices are of the same order.

Example 1

Calculate $\begin{pmatrix} 3 & 2 & 1 \\ 0 & -4 & 5 \end{pmatrix} + \begin{pmatrix} -1 & 1 & 3 \\ 2 & 4 & 0 \end{pmatrix}$

Solution

Both matrices are of the order 2×3. So we can add them. Add the corresponding elements.

$$\begin{pmatrix} 3+(-1) & 2+1 & 1+3 \\ 0+2 & -4+4 & 5+0 \end{pmatrix} = \begin{pmatrix} 2 & 3 & 4 \\ 2 & 0 & 5 \end{pmatrix}$$

Example 2

Calculate $\begin{pmatrix} 5 & 2 \\ 1 & -4 \end{pmatrix} + \begin{pmatrix} 3 & 2 & 1 \\ 0 & -4 & 5 \end{pmatrix}$

Solution

The two matrices are of different order so we cannot add them.

If A and B are matrices of the same order then:

- $A + B = B + A$
- $A + O = O + A$ (where O is the zero matrix of same order as A)

Subtraction of matrices

Subtraction is similar to addition. The matrices must have the same order. Subtract the elements of the second matrix from the corresponding elements of the first matrix.

Example 1

Calculate $\begin{pmatrix} 3 & 2 & 4 \\ 0 & -2 & 5 \end{pmatrix} - \begin{pmatrix} 5 & 2 & 1 \\ 1 & -4 & 5 \end{pmatrix}$

Solution

The matrices have the same order, so we can carry out the subtraction. Subtract the elements of the second matrix from the elements of the first matrix.

$$\begin{pmatrix} 3-5 & 2 \quad - \quad 2 & 4-1 \\ 0-1 & -2-(-4) & 5-5 \end{pmatrix} = \begin{pmatrix} -2 & 0 & 3 \\ -1 & 2 & 0 \end{pmatrix}$$

Example 2

Determine $\begin{pmatrix} 3 & 2 & 4 \\ 0 & -2 & 5 \end{pmatrix} - \begin{pmatrix} 5 & 2 \\ 1 & -4 \end{pmatrix}$

Solution

The two matrices have different orders which means they are not compatible for subtraction.

Scalar multiplication

At the beginning of this unit you saw Mary's weekly shopping list. She found this exercise times-consuming and demanding. So she decided to buy everything at the beginning of each month (four weeks). Find out how much of each item she has to buy.

It is not difficult. Simply add the matrices for the four weeks.

$$\begin{pmatrix} 2 & 2 & 4 \\ 10 & 8 & 12 \\ 3 & 2 & 3 \\ 1 & 0 & 1 \end{pmatrix} + \begin{pmatrix} 2 & 2 & 4 \\ 10 & 8 & 12 \\ 3 & 2 & 3 \\ 1 & 0 & 1 \end{pmatrix} + \begin{pmatrix} 2 & 2 & 4 \\ 10 & 8 & 12 \\ 3 & 2 & 3 \\ 1 & 0 & 1 \end{pmatrix} + \begin{pmatrix} 2 & 2 & 4 \\ 10 & 8 & 12 \\ 3 & 2 & 3 \\ 1 & 0 & 1 \end{pmatrix} = \begin{pmatrix} 8 & 8 & 16 \\ 40 & 32 & 48 \\ 12 & 8 & 12 \\ 4 & 0 & 4 \end{pmatrix}$$

This can be written as $4\begin{pmatrix} 2 & 2 & 4 \\ 10 & 8 & 12 \\ 3 & 2 & 3 \\ 1 & 0 & 1 \end{pmatrix}$. In other words, you can multiply a matrix by a number.

To multiply a matrix by a number, multiply each element by that number.

Example 1 and solution

$$3\begin{pmatrix} 6 & -1 \\ 3 & 2 \end{pmatrix} = \begin{pmatrix} 3\times 6 & 3\times(-1) \\ 3\times 3 & 3 \times 2 \end{pmatrix} = \begin{pmatrix} 18 & -3 \\ 9 & 6 \end{pmatrix}$$

Example 2 and solution

$$-2\begin{pmatrix} -5 & 0 \\ 2 & 3 \end{pmatrix} = \begin{pmatrix} -2 \times -5 & -2\times 0 \\ -2 \times 2 & -2\times 3 \end{pmatrix} = \begin{pmatrix} 10 & 0 \\ -4 & -6 \end{pmatrix}$$

Example 3 and solution

$$4\begin{pmatrix} 3 \times & 2y \\ -5 \times & y \end{pmatrix} = \begin{pmatrix} 4\times(3x) & 4\times(2y) \\ 4\times(-5x) & 4\times(y) \end{pmatrix} = \begin{pmatrix} 12x & 8y \\ -20x & 4y \end{pmatrix}$$

Example 4 and solution

$$5\begin{pmatrix} -2 \\ 1 \end{pmatrix} - 3\begin{pmatrix} 4 \\ -3 \end{pmatrix} = \begin{pmatrix} -10 \\ 5 \end{pmatrix} - \begin{pmatrix} 12 \\ -9 \end{pmatrix} = \begin{pmatrix} -10-12 \\ 5+9 \end{pmatrix} = \begin{pmatrix} -22 \\ 14 \end{pmatrix}$$

Now try some yourself.

EXERCISE 19

Find:

1. $3(3 \quad -1)$
2. $3\begin{pmatrix} 5 \\ -2 \end{pmatrix}$
3. $2\begin{pmatrix} 5 & -3 \\ 0 & 1 \end{pmatrix}$
4. $-4\begin{pmatrix} 1 & -2 \\ 3 & 0 \end{pmatrix}$
5. $\frac{1}{2}\begin{pmatrix} 4 & 0 \\ 2 & 1 \end{pmatrix}$
6. $3\begin{pmatrix} 2a & -2b \\ 3a & 4b \end{pmatrix}$
7. $-2\begin{pmatrix} 1 & 0 \\ 0 & -1 \end{pmatrix}$
8. $2\begin{pmatrix} 4 \\ -1 \end{pmatrix} - 3\begin{pmatrix} 1 \\ -2 \end{pmatrix}$

Check your answers at the end of this module.

Multiplication of matrices

John likes sweets and chocolates.

In the morning he buys 6 sweets and 4 bars of chocolate.
In the evening he buys only half these amounts.
One sweet costs 5 cents and one bar of chocolate costs 20 cents.

Find out how many cents he spends every day on sweets and chocolates.

	Sweets	Chocolate bars
Morning	6	4
Evening	3	2

Items	Price
Sweet	5 cents
Chocolate	20 cents

We can represent the above data in two matrices

$$\begin{pmatrix} 6 & 4 \\ 3 & 2 \end{pmatrix} \text{ and } \begin{pmatrix} 5 \\ 20 \end{pmatrix}$$

Money spent on sweets in the morning $= 6 \times 5 = 30$ cents
Money spent on chocolate in the morning $= 4 \times 20 = 80$ cents
Total money spent in the morning $= (30 + 80)$ cents
$= 110$ cents

$$\begin{pmatrix} 6 & 4 \\ - & \end{pmatrix} \begin{pmatrix} 5 \\ 20 \end{pmatrix}$$

Money spent on sweets in the evening $= 3 \times 5 = 15$ cents
Money spent on chocolate in the evening $= 2 \times 20 = 40$ cents
Total money spent in the evening $= (15 + 40)$ cents $= 55$ cents

$$\begin{pmatrix} - & - \\ 3 & 2 \end{pmatrix} \begin{pmatrix} 5 \\ 20 \end{pmatrix}$$

We can show the money spent as Morning 110 cents
Evening 55 cents

or as the matrix $\begin{pmatrix} 110 \\ 55 \end{pmatrix}$.

We represent the calculation as a multiplication of matrices

$$\begin{pmatrix} 6 & 4 \\ 3 & 2 \end{pmatrix}\begin{pmatrix} 5 \\ 20 \end{pmatrix} = \begin{pmatrix} 110 \\ 55 \end{pmatrix}$$

Using this example we can formulate the rules for multiplying matrices:

- The number of columns of the first matrix must be equal to the number of rows of the second matrix.
- The order of the product is the number of rows of the first matrix by the number of columns in the second matrix. For example, if the order of the first matrix is $m \times r$ and the order of the second matrix is $r \times n$ then the order of the product will be $m \times n$.
- Multiply the elements of a row of the first matrix and a column of the second matrix and add the products. For example, the sum of the products of the elements of the third row of the first matrix multiplied by the elements of the second column of the second matrix gives the element in the third row and second column of the product matrix.

This probably sounds quite confusing to you so follow the examples carefully and then try some on your own.

Example 1

Find the product $(2 \quad 5)\begin{pmatrix} 4 \\ 3 \end{pmatrix}$

Solution

Order of the first matrix: 1×2
Order of the second matrix: 2×1
The number of the columns of the first matrix = the number of rows of the second matrix = 2.
Therefore, multiplication is possible and the order of the product will be 1×1.
Multiply the corresponding elements and add them up.

That is:
multiply the first element of the row in the first matrix by the first element of the column in the second matrix: $2 \times 4 = 8$

multiply the second element of the row in the first matrix by the second element of the column in the second matrix: $5 \times 3 = 15$

and add them up, $8 + 15 = 23$

$$(2 \quad 5)\begin{pmatrix} 4 \\ 3 \end{pmatrix} = (23)$$

Example 2

Find the product $\begin{pmatrix} 2 & 3 \\ 1 & 0 \end{pmatrix}\begin{pmatrix} 3 & 1 \\ 6 & 4 \end{pmatrix}$

Solution

Orders of the matrices are 2×2 and 2×2. Multiplication is possible. Order of the product will be 2×2.

Multiply the corresponding elements of the first row of the first matrix by the elements in the first column of the second matrix and add the products.

$$(2 \ \ 3)\begin{pmatrix} 3 \\ 6 \end{pmatrix} = 2 \times 3 + 3 \times 6 = 6 + 18 = 24$$

This number is the element in the first row and first column of the answer $\begin{pmatrix} 24 & . \\ . & . \end{pmatrix}$

Now multiply the corresponding elements of the first row of the first matrix by the elements of the second column of the second matrix and add the results.

$$(2 \ \ 3)\begin{pmatrix} 1 \\ 4 \end{pmatrix} = 2 \times 1 + 3 \times 4 = 2 + 12 = 14$$

This number goes in the first row, second column of the answer $\begin{pmatrix} 24 & 14 \\ . & . \end{pmatrix}$

Next multiply the corresponding elements of the second row of the first matrix by the first column of the second matrix and add them up.

$$(1 \ \ 0)\begin{pmatrix} 3 \\ 6 \end{pmatrix} = 1 \times 3 + 0 \times 6 = 3 + 0 = 3$$

This number goes in the second row, first column of the answer $\begin{pmatrix} 24 & 14 \\ 3 & . \end{pmatrix}$

Finally, multiply the corresponding elements of the second row of the first matrix by the second column of the second matrix and add the results.

$$(1 \ \ 0)\begin{pmatrix} 1 \\ 4 \end{pmatrix} = 1 \times 1 + 0 \times 4 = 1 + 0 = 1$$

This is the element in the second row, second column of the answer $\begin{pmatrix} 24 & 14 \\ 3 & 1 \end{pmatrix}$

Example 3

Calculate $\begin{pmatrix} 4 & 5 \\ 3 & 1 \end{pmatrix}\begin{pmatrix} 4 & 2 & -2 \\ 3 & 1 & 0 \end{pmatrix}$

Solution

The orders of the matrices are 2×2 and 2×3. Therefore, the order of the product is 2×3.

$$\begin{pmatrix} 4 & 5 \\ 3 & 1 \end{pmatrix}\begin{pmatrix} 4 & 2 & -2 \\ 3 & 1 & 0 \end{pmatrix} = \begin{pmatrix} (4)(4)+(5)(3) & (4)(2)+(5)(1) & (4)(-2)+(5)(0) \\ (3)(4)+(1)(3) & (3)(2)+(1)(1) & (3)(-2)+(1)(0) \end{pmatrix}$$

$$= \begin{pmatrix} 16+15 & 8+5 & -8+0 \\ 12+3 & 6+1 & -6+0 \end{pmatrix} = \begin{pmatrix} 31 & 13 & -8 \\ 15 & 7 & -6 \end{pmatrix}$$

Example 4

Work out the matrix product $\begin{pmatrix} 3 & 1 & -2 \\ 5 & 0 & 4 \end{pmatrix}(-2\ 0\ 5)$.

Solution

The orders of the matrices are 2×3 and 1×3.

Since the number of columns in the first matrix is not equal to the number of rows in the second matrix, we cannot multiply the two matrices.

Example 5

Given that $\begin{pmatrix} x & 2 \\ 3 & y \end{pmatrix}\begin{pmatrix} 4 \\ -1 \end{pmatrix} = \begin{pmatrix} 8 \\ 5 \end{pmatrix}$, find the value of x and the value of y.

Solution

Multiplying the matrices on the left-hand side gives $\begin{pmatrix} 4x-2 \\ 12-y \end{pmatrix}$.

This matrix has to be equal to $\begin{pmatrix} 8 \\ 5 \end{pmatrix}$, so $4x - 2 = 8$ and $12 - y = 5$.

Hence, $4x = 10$ and $12 - 5 = y$
that is $x = 2.5$ and $y = 7$

Example 6

Given that $A = \begin{pmatrix} 1 & 2 \\ 3 & 4 \end{pmatrix}$, find A^2 and A^3.

Solution

As in ordinary algebra, A^2 is shorthand for $A \times A$
and A^3 is shorthand for $A \times A \times A$.

$$A^2 = \begin{pmatrix} 1 & 2 \\ 3 & 4 \end{pmatrix}\begin{pmatrix} 1 & 2 \\ 3 & 4 \end{pmatrix} = \begin{pmatrix} (1\times1)+(2\times3) & (1\times2)+(2\times4) \\ (3\times1)+(4\times3) & (3\times2)+(4\times4) \end{pmatrix} = \begin{pmatrix} 7 & 10 \\ 15 & 22 \end{pmatrix}$$

$$A^3 = A \times A \times A = A^2 \times A = \begin{pmatrix} 7 & 10 \\ 15 & 22 \end{pmatrix}\begin{pmatrix} 1 & 2 \\ 3 & 4 \end{pmatrix}$$

$$= \begin{pmatrix} (7 \times 1) + (10 \times 3) & (7 \times 2) + (10 \times 4) \\ (15 \times 1) + (22 \times 3) & (15 \times 2) + (22 \times 4) \end{pmatrix}$$

$$= \begin{pmatrix} 37 & 54 \\ 81 & 118 \end{pmatrix}$$

Note: If you take A^3 to be $A \times A^2$, you will get the same answer. Try it yourself.

Multiplying matrices is quite tricky and you have to be very careful. I know you'll need to go over the examples again (try doing them yourself) before you tackle the next exercise.

EXERCISE 20

1. Find the following matrix products.

a) $\begin{pmatrix} 4 & 5 \end{pmatrix}\begin{pmatrix} 3 \\ 2 \end{pmatrix}$ b) $\begin{pmatrix} -2 & 1 \end{pmatrix}\begin{pmatrix} -3 \\ 0 \end{pmatrix}$

c) $\begin{pmatrix} -3 & -2 \end{pmatrix}\begin{pmatrix} 3 \\ 4 \end{pmatrix}$ d) $\begin{pmatrix} 3 & -2 \end{pmatrix}\begin{pmatrix} -2 \\ 3 \end{pmatrix}$

e) $\begin{pmatrix} a & 3 \end{pmatrix}\begin{pmatrix} 1 \\ 3 \end{pmatrix}$

2. Carry out the following matrix multiplications.

a) $\begin{pmatrix} 5 & 1 \\ 2 & 0 \end{pmatrix}\begin{pmatrix} 2 \\ 3 \end{pmatrix}$

b) $\begin{pmatrix} -1 & -2 \\ -3 & 0 \end{pmatrix}\begin{pmatrix} -2 \\ -3 \end{pmatrix}$

c) $\begin{pmatrix} 2 & 6 \\ 0 & 3 \end{pmatrix}\begin{pmatrix} 1 & 3 \\ 1 & 4 \end{pmatrix}$

d) $\begin{pmatrix} 1 & -2 \\ 3 & 5 \end{pmatrix}\begin{pmatrix} 3 & -1 \\ -1 & 1 \end{pmatrix}$

3. Find the value of x in each of the following matrix equations.

a) $\begin{pmatrix} x & 2 \end{pmatrix}\begin{pmatrix} 1 \\ 3 \end{pmatrix} = (10)$

b) $\begin{pmatrix} 3 & x \end{pmatrix}\begin{pmatrix} 5 \\ 2 \end{pmatrix} = (17)$

c) $\begin{pmatrix} x & -5 \end{pmatrix}\begin{pmatrix} 2 \\ 3 \end{pmatrix} = (-1)$

4. If $A = \begin{pmatrix} 0 & 5 \\ 2 & -2 \end{pmatrix}$, $B = \begin{pmatrix} 1 & 3 \\ -1 & 2 \end{pmatrix}$ and $I = \begin{pmatrix} 1 & 0 \\ 0 & 1 \end{pmatrix}$, find

a) AB b) BA
c) AI d) IA
e) B^3

Check your answers at the end of this module.

Multiplication by a unit matrix

Multiplying ordinary numbers by one leaves the numbers unchanged.

For example, $4 \times 1 = 4$, $1 \times 150 = 150$, $a \times 1 = a$.

Similarly, a square matrix is unaltered by multiplication by a unit matrix.

$AI = IA = A$ (where 1 is a unit matrix of the same order as A).

Order of matrices in multiplication

If A and B are two matrices, then AB is *not* in general equal to BA. That is, multiplication of matrices in general is not commutative. To avoid confusion in the multiplication of matrices, AB may be described as A post-multiplied by B, or B pre-multiplied by A. BA is B post-multiplied by A or A pre-multiplied by B.

Determinant of a matrix

If $A = \begin{pmatrix} 2 & 1 \\ 3 & 5 \end{pmatrix}$, then the number $(2 \times 5) - (1 \times 3)$ is called the determinant of the matrix A. In this particular case, the determinant is 7.

The determinant of A is denoted by det A or $|A|$ or $\triangle$.
We will be using $|A|$.

> If $A = \begin{pmatrix} a & b \\ c & d \end{pmatrix}$, $|A| = ad - bc$.
>
> In words, $|A|$ = the product of the elements in the leading diagonal – the product of the elements in the other diagonal.

Example 1

If $M = \begin{pmatrix} 5 & -2 \\ 1 & 3 \end{pmatrix}$, find the value of $|M|$.

Solution

Product of the elements in the leading diagonal $= 5 \times 3 = 15$.
Product of the elements in the other diagonal $= 1 \times (-2) = -2$.

$|M|$ = product of the elements in the leading diagonal – product of the elements in the other diagonal.

$|M| = 15 - (-2) = 15 + 2 = 17$

Example 2

Find the value of x if the determinant of $\begin{pmatrix} x & 3 \\ 4 & 1 \end{pmatrix} = 2$.

Solution

The determinant of the matrix $= (x \times 1) - (3 \times 4) = x - 12$. This has to be equal to 2.

So $x - 12 = 2$
$x = 14$

The following exercise concentrates on the idea of the determinant of a matrix. It should give you less trouble than Exercise 20!

EXERCISE 21

1. Find the determinants of the following matrices.

a) $\begin{pmatrix} 4 & 1 \\ 3 & 5 \end{pmatrix}$ b) $\begin{pmatrix} 6 & 8 \\ 3 & 4 \end{pmatrix}$

c) $\begin{pmatrix} 2 & -1 \\ 3 & 9 \end{pmatrix}$ d) $\begin{pmatrix} -1 & 0 \\ 0 & 1 \end{pmatrix}$

e) $\begin{pmatrix} -3 & -2 \\ -1 & 1 \end{pmatrix}$

2. If $B = \begin{pmatrix} x & 2 \\ 3 & 1 \end{pmatrix}$ and $|B| = 4$, find the value of x.

3. Given that $C = \begin{pmatrix} 1 & 4 \\ -2 & x \end{pmatrix}$ and $|C| = 5$, find the value of x.

4. Find y if the determinant of the matrix $\begin{pmatrix} y & 2 \\ 3 & 2 \end{pmatrix} = 2$.

5. Given that $A = \begin{pmatrix} -2 & p \\ -2 & 3 \end{pmatrix}$ and $|A| = 12$, find the value of p.

Have you grasped the idea of the determinant of a matrix? Check your answers at the end of this module.

The inverse of a matrix

You should already know that $5 \times \frac{1}{5} = 1$ and $5 \times 5^{-1} = 1$.

Here, we say that 5^{-1} (or $\frac{1}{5}$) is the multiplicative inverse of 5 and 5 is the multiplicative inverse of 5^{-1} (or $\frac{1}{5}$).

> A number multiplied by its multiplicative inverse is one.

The multiplicative inverse of 3 is 3^{-1}.
The multiplicative inverse of -2 is $(-2)^{-1}$
a is the multiplicative inverse of a^{-1}.

The inverse of a square matrix A is denoted by A^{-1} and $A.A^{-1} = A^{-1}.A = I$, where I is the unit matrix of the same order as A.

If $A = \begin{pmatrix} 5 & 2 \\ 7 & 3 \end{pmatrix}$ and $B = \begin{pmatrix} 3 & -2 \\ -7 & 5 \end{pmatrix}$,

then $\begin{pmatrix} 5 & 2 \\ 7 & 3 \end{pmatrix}\begin{pmatrix} 3 & -2 \\ -7 & 5 \end{pmatrix} = \begin{pmatrix} 3 & -2 \\ -7 & 5 \end{pmatrix}\begin{pmatrix} 5 & 2 \\ 7 & 3 \end{pmatrix} = \begin{pmatrix} 1 & 0 \\ 0 & 1 \end{pmatrix}$.

Therefore, B is the inverse of A and A is the inverse of B.

Look at the following multiplications:

$$\begin{pmatrix} 2 & 3 \\ 5 & 8 \end{pmatrix}\begin{pmatrix} 8 & -3 \\ -5 & 2 \end{pmatrix} = \begin{pmatrix} 1 & 0 \\ 0 & 1 \end{pmatrix}$$

$$\begin{pmatrix} 14 & 1 \\ 13 & 1 \end{pmatrix}\begin{pmatrix} 1 & -1 \\ -13 & 14 \end{pmatrix} = \begin{pmatrix} 1 & 0 \\ 0 & 1 \end{pmatrix}$$

$$\begin{pmatrix} 3 & 1 \\ 11 & 4 \end{pmatrix}\begin{pmatrix} 4 & -1 \\ -11 & 3 \end{pmatrix} = \begin{pmatrix} 1 & 0 \\ 0 & 1 \end{pmatrix}$$

In the above multiplications the second matrix is the inverse of the first matrix because the product is a unit matrix.

Did you notice how the second matrix resembles the first matrix?

- The elements of the leading diagonal are swopped.
- The elements of the other diagonal are changed in sign.

We can use the above rule to find the inverse of a matrix.

EXERCISE 22

Find the inverses of the following matrices.

1. $\begin{pmatrix} 3 & 2 \\ 1 & 1 \end{pmatrix}$
2. $\begin{pmatrix} 1 & 1 \\ 1 & 2 \end{pmatrix}$
3. $\begin{pmatrix} 3 & 4 \\ 5 & 7 \end{pmatrix}$
4. $\begin{pmatrix} 5 & -7 \\ -2 & 3 \end{pmatrix}$
5. $\begin{pmatrix} 3 & -2 \\ -4 & 3 \end{pmatrix}$

You should have checked your answers by using $A.A^{-1} = \begin{pmatrix} 1 & 0 \\ 0 & 1 \end{pmatrix}$.
If you have done this, you will know which (if any) of your answers are wrong. The correct answers are given at the end of this module.

Example 1

Find the inverse of the matrix $\begin{pmatrix} 3 & 2 \\ 5 & 4 \end{pmatrix}$.

Solution

By our rule, the inverse is $\begin{pmatrix} 4 & -2 \\ -5 & 3 \end{pmatrix}$

But this answer is not correct because

$$\begin{pmatrix} 3 & 2 \\ 5 & 4 \end{pmatrix}\begin{pmatrix} 4 & -2 \\ -5 & 3 \end{pmatrix} = \begin{pmatrix} 4 & -2 \\ -5 & 3 \end{pmatrix}\begin{pmatrix} 3 & 2 \\ 5 & 4 \end{pmatrix} = \begin{pmatrix} 2 & 0 \\ 0 & 2 \end{pmatrix}.$$

This is not a unit matrix.

$$\begin{pmatrix} 2 & 0 \\ 0 & 2 \end{pmatrix} = 2\begin{pmatrix} 1 & 0 \\ 0 & 1 \end{pmatrix}$$

Hence, we must divide either the first or the second matrix by 2 in order to obtain the unit matrix.

$$\frac{1}{2}\begin{pmatrix} 3 & 2 \\ 5 & 4 \end{pmatrix}\begin{pmatrix} 4 & -2 \\ -5 & 3 \end{pmatrix} = \begin{pmatrix} 1 & 0 \\ 0 & 1 \end{pmatrix}$$

Therefore, the inverse of $\begin{pmatrix} 3 & 2 \\ 5 & 4 \end{pmatrix}$ is $\frac{1}{2}\begin{pmatrix} 4 & -2 \\ -5 & 3 \end{pmatrix}$ or $\begin{pmatrix} \frac{4}{2} & -\frac{2}{2} \\ -\frac{5}{2} & \frac{3}{2} \end{pmatrix}$.

To obtain the unit matrix, we have to divide by a particular number. To find this number it is not necessary to multiply the two matrices.

The number is actually the determinant of the matrix, $ad - bc$.

In general terms, $\begin{pmatrix} a & b \\ c & d \end{pmatrix}\begin{pmatrix} d & -b \\ -c & a \end{pmatrix} = \begin{pmatrix} ad-bc & 0 \\ 0 & ad-bc \end{pmatrix}$

$$= (ad - bc)\begin{pmatrix} 1 & 0 \\ 0 & 1 \end{pmatrix}.$$

Hence the inverse of $\begin{pmatrix} a & b \\ c & d \end{pmatrix} = \frac{1}{ad-bc}\begin{pmatrix} d & -b \\ -c & a \end{pmatrix}$.

So to find the inverse of a matrix:

- Calculate the determinant of the matrix.
- If the determinant is 1, there is no problem. By our rule, swop the elements of the leading diagonal and change the signs of the elements of the other diagonal.
- If the determinant is not 1 or 0, swop the elements of the leading diagonal, change the signs of the elements of the other diagonal, and finally divide each element by the determinant.
- If the determinant is 0, the method breaks down because division by 0 is impossible. In such a case, the matrix has no inverse and it is called a singular matrix.

Example 1

Find the inverse of $\begin{pmatrix} 2 & 1 \\ 4 & 2 \end{pmatrix}$.

Solution

The determinant = the product of the elements in the leading diagonal − the product of the elements in the other diagonal
$= (2 \times 2) - (4 \times 1) = 4 - 4 = 0$.
Therefore, the matrix is a singular matrix and the inverse cannot be found.

Example 2

If $A = \begin{pmatrix} 8 & 5 \\ 3 & 2 \end{pmatrix}$, find A^{-1}.

Solution

$|A| = (8 \times 2) - (3 \times 5) = 16 - 15 = 1$

$$A^{-1} = \begin{pmatrix} 2 & -5 \\ -3 & 8 \end{pmatrix}$$

Example 3

Given that $A = \begin{pmatrix} 5 & 7 \\ 6 & 9 \end{pmatrix}$, find A^{-1}.

Solution

$|A| = (5 \times 9) - (6 \times 7) = 45 - 42 = 3$

$$A^{-1} = \frac{1}{3}\begin{pmatrix} 9 & -7 \\ -6 & 5 \end{pmatrix} = \begin{pmatrix} \frac{9}{3} & -\frac{7}{3} \\ -\frac{6}{3} & \frac{5}{3} \end{pmatrix} = \begin{pmatrix} 3 & -\frac{7}{3} \\ -2 & \frac{5}{3} \end{pmatrix}$$

Example 4

Given that $\begin{pmatrix} 8 & 5 \\ 3 & 2 \end{pmatrix}\begin{pmatrix} x \\ y \end{pmatrix} = \begin{pmatrix} 14 \\ 5 \end{pmatrix}$, find the value of x and the value of y.

Solution

From Example 2, the inverse of $\begin{pmatrix} 8 & 5 \\ 3 & 2 \end{pmatrix}$ is $\begin{pmatrix} 2 & -5 \\ -3 & 8 \end{pmatrix}$.

Multiplying both sides of the given equation by this inverse matrix gives $\begin{pmatrix} 2 & -5 \\ -3 & 8 \end{pmatrix}\begin{pmatrix} 8 & 5 \\ 3 & 2 \end{pmatrix}\begin{pmatrix} x \\ y \end{pmatrix} = \begin{pmatrix} 2 & -5 \\ -3 & 8 \end{pmatrix}\begin{pmatrix} 14 \\ 5 \end{pmatrix}$

note that on both sides of the equation the inverse matrix is on the left

Working out the products gives $\begin{pmatrix} 1 & 0 \\ 0 & 1 \end{pmatrix} \begin{pmatrix} x \\ y \end{pmatrix} = \begin{pmatrix} 3 \\ -2 \end{pmatrix}$

This is $\begin{pmatrix} x \\ y \end{pmatrix} = \begin{pmatrix} 3 \\ -2 \end{pmatrix}$

Hence, $x = 3$ and $y = -2$.

Note: The given equation is equivalent to the simultaneous equations:
$8x + 5y = 14$
$3x + 2y = 5$

You could have found the value of x and y by solving these simultaneous equations. Using inverse matrices gives us another method of solving linear simultaneous equations.

Finding the inverse of a matrix is usually part of a longer problem such as solving simultaneous equations or dealing with geometrical transformations (as you will see later).

Here is some practice for you on finding inverses.

EXERCISE 23

1. State whether each of the following matrices has an inverse. If the inverse exists, find it.

a) $\begin{pmatrix} 4 & 6 \\ 1 & 2 \end{pmatrix}$ b) $\begin{pmatrix} 2 & 3 \\ 5 & 8 \end{pmatrix}$

c) $\begin{pmatrix} 3 & 2 \\ 6 & 4 \end{pmatrix}$ d) $\begin{pmatrix} 3 & -2 \\ 2 & -1 \end{pmatrix}$

e) $\begin{pmatrix} 3 & 4 \\ 5 & 6 \end{pmatrix}$

2. Show that I is its own inverse, where $I = \begin{pmatrix} 1 & 0 \\ 0 & 1 \end{pmatrix}$.

3. Show that each of the following matrices is its own inverse.

a) $\begin{pmatrix} 1 & 0 \\ 0 & -1 \end{pmatrix}$ b) $\begin{pmatrix} -1 & 0 \\ 0 & -1 \end{pmatrix}$

c) $\begin{pmatrix} 0 & 1 \\ 1 & 0 \end{pmatrix}$ d) $\begin{pmatrix} 0 & -1 \\ -1 & 0 \end{pmatrix}$

4. a) Given that $A = \begin{pmatrix} 3 & 7 \\ 2 & 5 \end{pmatrix}$, find A^{-1}.

b) Use a matrix method to solve the simultaneous equations:
$3x + 7y = 5$
$2x + 5y = 3$

Check your answers at the end of this module.

C Using matrices for geometrical transformations

Matrices can be used to describe geometrical transformations in a plane.

Suppose P is the point (4, 3). That is, P is the point whose position vector is $\begin{pmatrix} 4 \\ 3 \end{pmatrix}$.

Let us pre-multiply this $\begin{pmatrix} 4 \\ 3 \end{pmatrix}$ matrix by the matrix $\begin{pmatrix} 2 & 0 \\ 0 & 2 \end{pmatrix}$.

$$\begin{pmatrix} 2 & 0 \\ 0 & 2 \end{pmatrix}\begin{pmatrix} 4 \\ 3 \end{pmatrix} = \begin{pmatrix} 8 \\ 6 \end{pmatrix}$$

The above multiplication results in another position vector $\begin{pmatrix} 8 \\ 6 \end{pmatrix}$, i.e. another point P′(8,6).

So the matrix $\begin{pmatrix} 2 & 0 \\ 0 & 2 \end{pmatrix}$ transforms the point P(4, 3) into P′(8, 6).

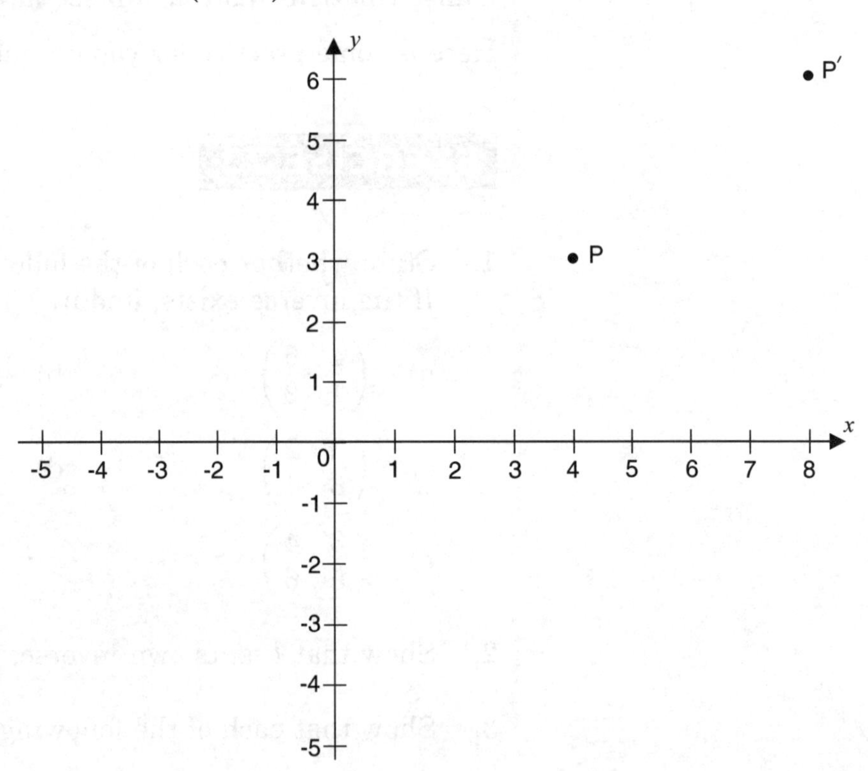

We can use this principle to describe all the transformations that we came across in Unit 3.

Reflection

Let me show you how we can find out the matrix that represents the reflection in the x-axis.

Consider the unit square, where O, A, B and C are the points (0, 0), (1, 0), (1, 1) and (0, 1).

Under the transformation reflection in the x-axis, the points O, A, B and C are transformed into O, A, B′ and C′.
The points B′ and C′ are (1, –1) and (0, –1).

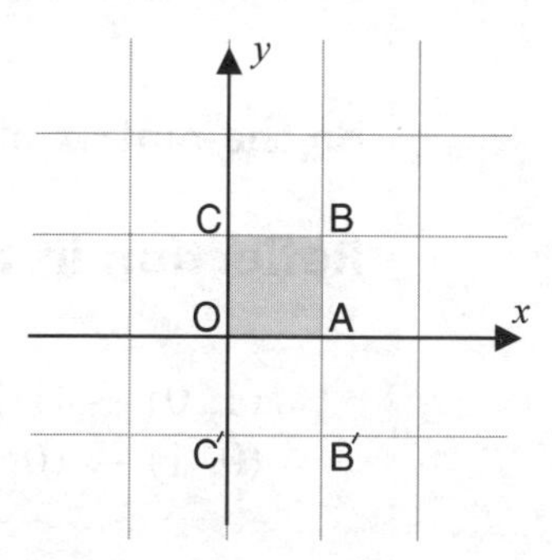

Let us assume the matrix $\begin{pmatrix} a & b \\ c & d \end{pmatrix}$ represents the transformation reflection in the x-axis.

Then $\begin{pmatrix} a & b \\ c & d \end{pmatrix}\begin{pmatrix} 0 \\ 0 \end{pmatrix} = \begin{pmatrix} 0 \\ 0 \end{pmatrix}$, $\begin{pmatrix} a & b \\ c & d \end{pmatrix}\begin{pmatrix} 1 \\ 0 \end{pmatrix} = \begin{pmatrix} 1 \\ 0 \end{pmatrix}$,

$\begin{pmatrix} a & b \\ c & d \end{pmatrix}\begin{pmatrix} 1 \\ 1 \end{pmatrix} = \begin{pmatrix} 1 \\ -1 \end{pmatrix}$ and $\begin{pmatrix} a & b \\ c & d \end{pmatrix}\begin{pmatrix} 0 \\ 1 \end{pmatrix} = \begin{pmatrix} 0 \\ -1 \end{pmatrix}$.

This can be written as

$$\begin{array}{c} \\ \begin{pmatrix} a & b \\ c & d \end{pmatrix} \end{array} \begin{array}{c} \text{O A B C} \\ \begin{pmatrix} 0 & 1 & 1 & 0 \\ 0 & 0 & 1 & 1 \end{pmatrix} \end{array} \begin{array}{c} = \\ = \end{array} \begin{array}{c} \text{O A B}' \text{ C}' \\ \begin{pmatrix} 0 & 1 & 1 & 0 \\ 0 & 0 & -1 & -1 \end{pmatrix} \end{array}$$

From this we get $a = 1$, $b = 0$, $c = 0$ and $d = -1$

so $\begin{pmatrix} a & b \\ c & d \end{pmatrix} = \begin{pmatrix} 1 & 0 \\ 0 & -1 \end{pmatrix}$

> make sure you can see where these values for a,b,c and d came from by multiplying the two matrices on the left and comparing the result with the matrix on the right

The matrix $\begin{pmatrix} 1 & 0 \\ 0 & -1 \end{pmatrix}$ represents a reflection in the x-axis.

There is a short cut to find out the matrix that represents a transformation.

> Select the points (1, 0) and (0, 1). Find the images of these points under the given transformation. Write down the position vectors of these images. The position vector of the image of the first point becomes the first column and the position vector of the image of the second point is the second column of the required matrix.

Reflection in the *x*-axis

$(1, 0) \to (1, 0)$
$(0, 1) \to (0, -1)$

The position vectors of these points are $\begin{pmatrix} 1 \\ 0 \end{pmatrix}$ and $\begin{pmatrix} 0 \\ -1 \end{pmatrix}$.

So the matrix of the transformation is $\begin{pmatrix} 1 & 0 \\ 0 & -1 \end{pmatrix}$.

Reflection in the *y*-axis

$(1, 0) \to (-1, 0)$
$(0, 1) \to (0, 1)$

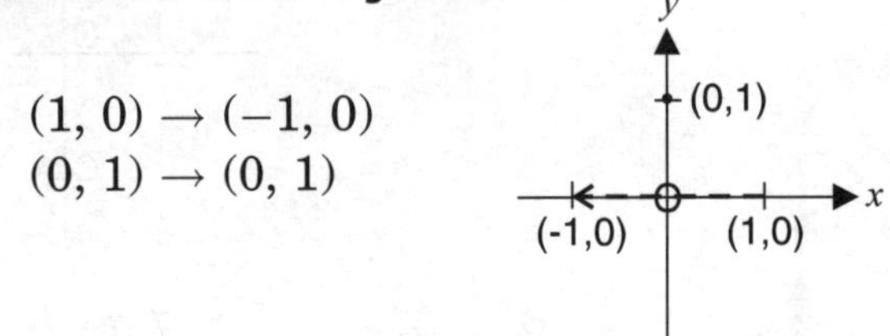

The position vectors of these points are $\begin{pmatrix} -1 \\ 0 \end{pmatrix}$ and $\begin{pmatrix} 0 \\ 1 \end{pmatrix}$.

So the matrix of the transformation is $\begin{pmatrix} -1 & 0 \\ 0 & 1 \end{pmatrix}$.

Reflection in the line $y = x$

$(1, 0) \to (0, 1)$
$(0, 1) \to (1, 0)$

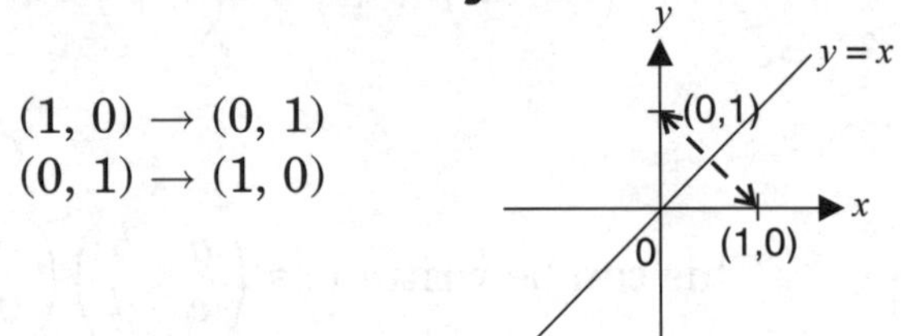

The position vectors of these points are $\begin{pmatrix} 0 \\ 1 \end{pmatrix}$ and $\begin{pmatrix} 1 \\ 0 \end{pmatrix}$.

So the matrix of the transformation is $\begin{pmatrix} 0 & 1 \\ 1 & 0 \end{pmatrix}$.

Reflection in the line $y = -x$

$(1, 0) \to (0, -1)$
$(0, 1) \to (-1, 0)$

The position vectors of these points are $\begin{pmatrix} 0 \\ -1 \end{pmatrix}$ and $\begin{pmatrix} -1 \\ 0 \end{pmatrix}$.

So the matrix of the transformation is $\begin{pmatrix} 0 & -1 \\ -1 & 0 \end{pmatrix}$.

Rotation about the origin through 90° anticlockwise (+90°)

$(1, 0) \rightarrow (0, 1)$
$(0, 1) \rightarrow (-1, 0)$

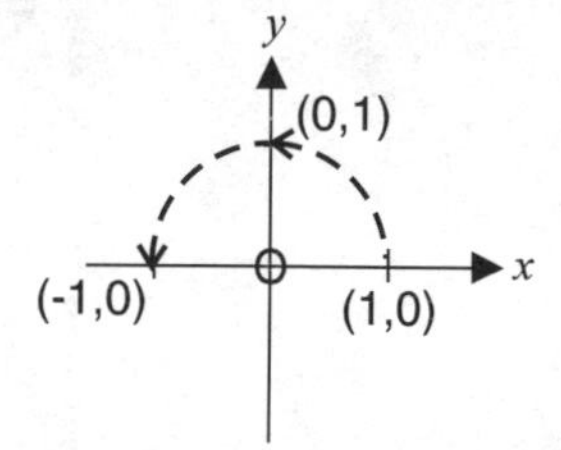

The transformation matrix is $\begin{pmatrix} 0 & -1 \\ 1 & 0 \end{pmatrix}$.

Rotation about the origin through 180° (half turn)

$(1, 0) \rightarrow (-1, 0)$
$(0, 1) \rightarrow (0, -1)$

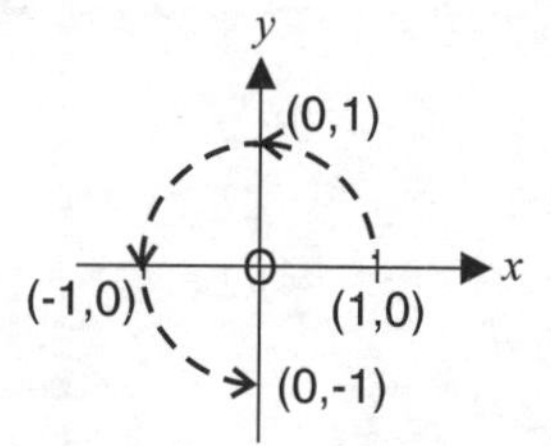

The transformation matrix is $\begin{pmatrix} -1 & 0 \\ 0 & -1 \end{pmatrix}$.

Rotation about the origin through 270° anticlockwise (+270°) or 90° clockwise (−90°)

$(1, 0) \rightarrow (0, -1)$
$(0, 1) \rightarrow (1, 0)$

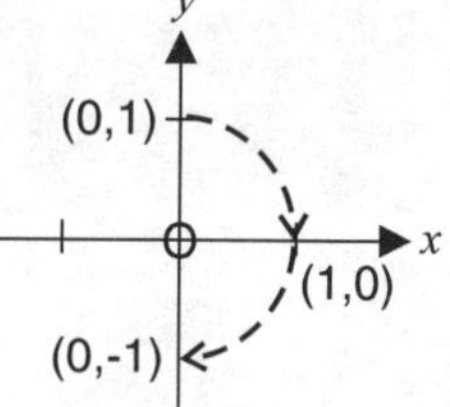

The transformation matrix is $\begin{pmatrix} 0 & 1 \\ -1 & 0 \end{pmatrix}$.

Stretch

One-way stretch parallel to the *x*-axis, with scale factor *k*, with the *y*-axis invariant

$(1, 0) \rightarrow (k, 0)$
$(0, 1) \rightarrow (0, 1)$

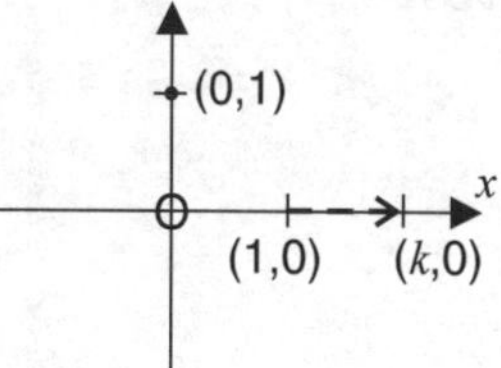

The transformation matrix is $\begin{pmatrix} k & 0 \\ 0 & 1 \end{pmatrix}$.

One-way stretch parallel to the *y*-axis, with scale factor *k*, with the *x*-axis invariant

$(1, 0) \rightarrow (1, 0)$
$(0, 1) \rightarrow (0, k)$

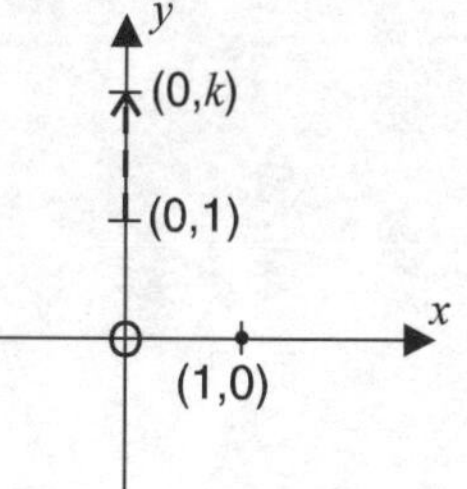

The transformation matrix is $\begin{pmatrix} 1 & 0 \\ 0 & k \end{pmatrix}$.

Enlargement

Enlargement with scale factor *k*, and centre of enlargement (0, 0)

$(1, 0) \to (k, 0)$
$(0, 1) \to (0, k)$

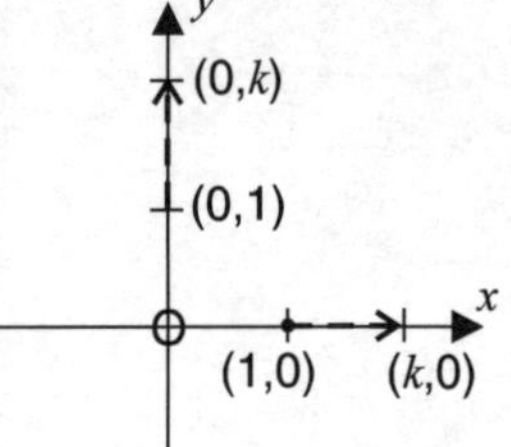

The transformation matrix is $\begin{pmatrix} k & 0 \\ 0 & k \end{pmatrix}$.

Shear

Shear in *x*-direction, *x*-axis invariant and scale factor *k*

$(1, 0) \to (1, 0)$
$(0, 1) \to (k, 1)$

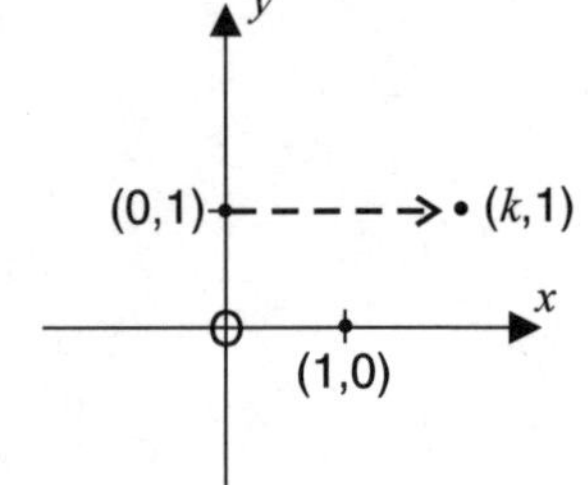

The transformation matrix is $\begin{pmatrix} 1 & k \\ 0 & 1 \end{pmatrix}$.

Shear in *y*-direction, *y*-axis invariant and scale factor *k*

$(1, 0) \to (1, k)$
$(0, 1) \to (0, 1)$

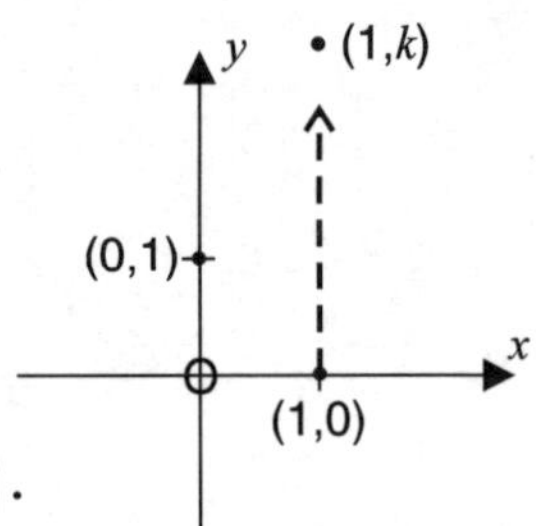

The transformation matrix is $\begin{pmatrix} 1 & 0 \\ k & 1 \end{pmatrix}$.

Translation

All the previous transformations are represented by a 2 by 2 matrix and, to find the image of a point, we have to pre-multiply the position vector $\begin{pmatrix} x \\ y \end{pmatrix}$ of the point by the appropriate matrix.

It follows that the image of the origin (0, 0) is the origin – in other words, the origin is invariant. This is not the case for a translation – every point moves, as there is no invariant point.

Translation is represented by a column vector or a column matrix of the order 2×1.

To obtain the image of a point under the translation we do not multiply by the matrix but we add the vector of the translation to the position vector $\begin{pmatrix} x \\ y \end{pmatrix}$ of the point.

Example 1

The figure shows the triangle PQR.

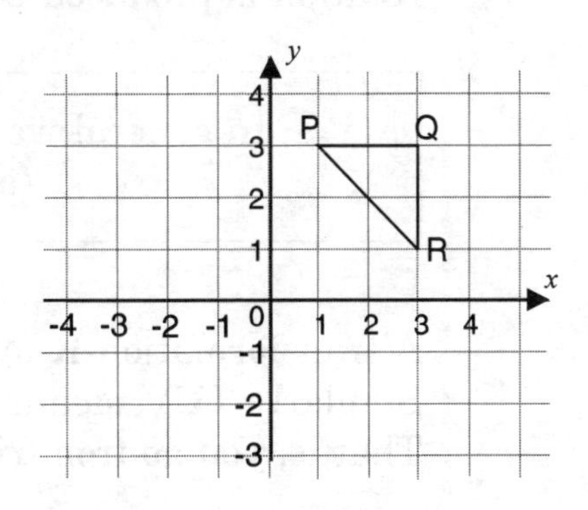

Show the new position of the triangle after the translation $\begin{pmatrix} -3 \\ +1 \end{pmatrix}$.

Solution

The position vector of P is $\begin{pmatrix} 1 \\ 3 \end{pmatrix}$.

The position vector of Q is $\begin{pmatrix} 3 \\ 3 \end{pmatrix}$.

The position vector of R is $\begin{pmatrix} 3 \\ 1 \end{pmatrix}$.

The image of P after the translation is $\begin{pmatrix} 1 \\ 3 \end{pmatrix} + \begin{pmatrix} -3 \\ +1 \end{pmatrix} = \begin{pmatrix} 1+(-3) \\ 3 \;+\; 1 \end{pmatrix}$

$= \begin{pmatrix} -2 \\ 4 \end{pmatrix}$

The image of Q after the translation is $\begin{pmatrix} 3 \\ 3 \end{pmatrix} + \begin{pmatrix} -3 \\ +1 \end{pmatrix} = \begin{pmatrix} 3+(-3) \\ 3 \;+\; 1 \end{pmatrix}$

$= \begin{pmatrix} 0 \\ 4 \end{pmatrix}$

The image of R after the translation is $\begin{pmatrix} 3 \\ 1 \end{pmatrix} + \begin{pmatrix} -3 \\ +1 \end{pmatrix} = \begin{pmatrix} 3+(-3) \\ 3 \;+\; 1 \end{pmatrix}$

$= \begin{pmatrix} 0 \\ 2 \end{pmatrix}$

The new position of PQR is P′Q′R′.

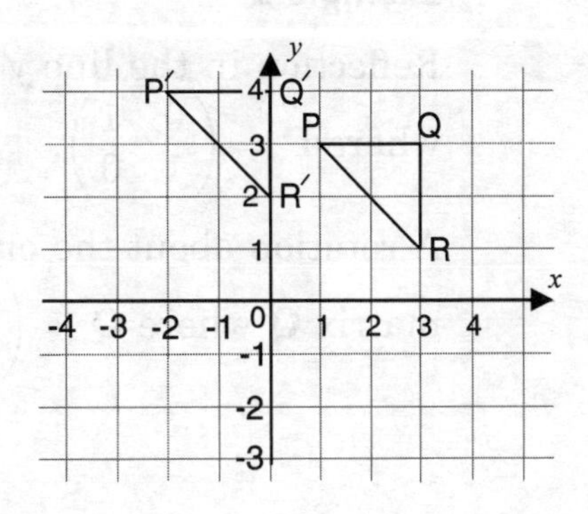

Notation

We normally denote a transformation by a capital letter. For example, the letter R can be used for a rotation. But when this rotation is produced by a matrix, then we use **R** (bold capital).

> In general we use a bold capital letter to represent a matrix transformation.

Successive transformations

A transformation **R** followed by a transformation **T** is denoted **TR**. Similarly **TUV** means **V** then **U** then **T**.
That is, you go from right to left.

If a figure PQRS is transformed using a matrix **A**, and the image is further transformed using a matrix **B**, then the final result is equivalent to a single transformation using the matrix **BA**.

Example 1

Reflection in the x-axis is represented by the matrix **A**,
where $\mathbf{A} = \begin{pmatrix} 1 & 0 \\ 0 & -1 \end{pmatrix}$.

Reflection in the y-axis is represented by the matrix **B**,
where $\mathbf{B} = \begin{pmatrix} -1 & 0 \\ 0 & 1 \end{pmatrix}$.

If a figure is reflected in the x-axis and then the image is reflected in the y-axis, the combined transformation is represented by the matrix **BA**.

$$\mathbf{BA} = \begin{pmatrix} -1 & 0 \\ 0 & 1 \end{pmatrix} \begin{pmatrix} 1 & 0 \\ 0 & -1 \end{pmatrix} = \begin{pmatrix} -1 & 0 \\ 0 & -1 \end{pmatrix}$$

You may recognise this matrix. It represents rotation about the origin through 180°.

This means that reflection in the x-axis followed by reflection in the y-axis is equivalent to a half-turn about the origin.

> you should convince yourself that this is right by reflecting a simple figure in the axes

Example 2

Reflection in the line $y = x$ is represented by the matrix **P**,
where $\mathbf{P} = \begin{pmatrix} 0 & 1 \\ 1 & 0 \end{pmatrix}$.

A rotation about the origin through +90° is represented by the matrix **Q** where $\mathbf{Q} = \begin{pmatrix} 0 & -1 \\ 1 & 0 \end{pmatrix}$.

If a figure is reflected in the line $y = x$ and then the image is rotated about the origin through $+90^{\circ}$, the combined transformation is represented by the matrix **QP**.

$$\mathbf{QP} = \begin{pmatrix} 0 & -1 \\ 1 & 0 \end{pmatrix} \begin{pmatrix} 0 & 1 \\ 1 & 0 \end{pmatrix} = \begin{pmatrix} -1 & 0 \\ 0 & 1 \end{pmatrix}$$

You may recognise this matrix. It represents reflection in the y-axis.

This means that reflection in the line $y = x$ followed by rotation about the origin through 90° anticlockwise is equivalent to reflection in the y-axis.

once again, you should convince yourself that this is true by reflecting and then rotating a simple figure

Note: The matrix **PQ** is not equal to the matrix **QP**.
Rotation about the origin through $+90^{\circ}$ followed by reflection in the line $y = x$ is *not* equivalent to reflection in the y-axis. What single transformation is equivalent to this combination?

Inverse transformations

The inverse of a transformation is such that it reverses the effect of that transformation.

If **L** transforms a figure ABCD onto A′B′C′D′, then $\mathbf{L}^{-1}$ (the inverse of **L**) will transform A′B′C′D′ back into ABCD.

Example 1

Consider a clockwise rotation of 90° about the origin (0, 0).
This is represented by the matrix **R**,
where $\mathbf{R} = \begin{pmatrix} 0 & 1 \\ -1 & 0 \end{pmatrix}$.

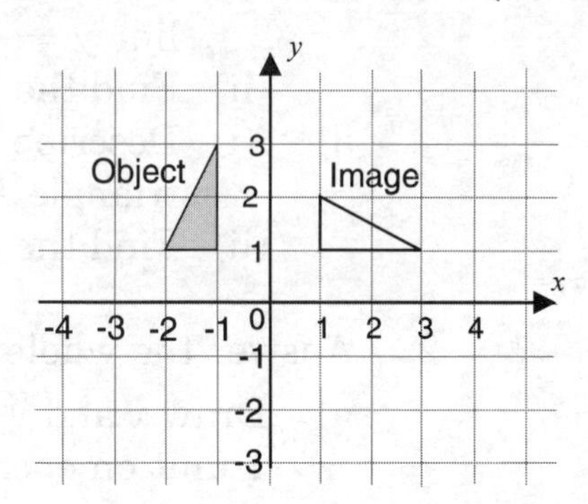

The inverse of **R** is $\mathbf{R}^{-1}$, where $\mathbf{R}^{-1} = \begin{pmatrix} 0 & -1 \\ 1 & 0 \end{pmatrix}$.

You will recognise this as the matrix which represents an anticlockwise rotation of 90° about the point (0, 0).

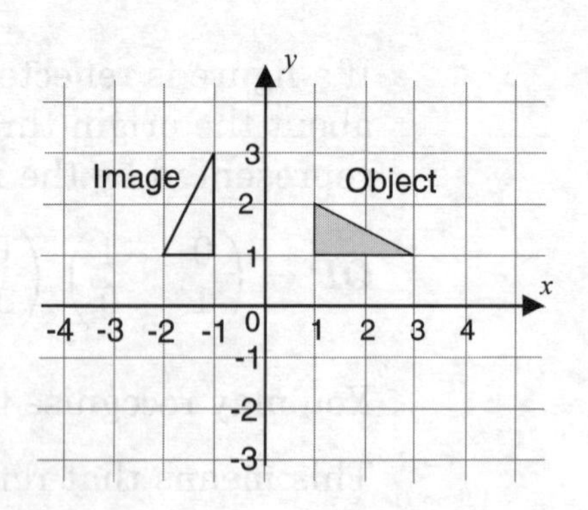

Clearly, an anticlockwise rotation of 90° about the point (0, 0) reverses the effect of a clockwise rotation of 90° about the point (0, 0).

Here are two questions for you to solve which combine the ideas of matrices and geometrical transformations.

EXERCISE 24

1. Answer the whole of this question on a sheet of graph paper.
   a) (i) Draw x and y axes from -6 to $+6$, using 1 cm to represent 1 unit of x and y.
      (ii) Draw the triangle ABC with vertices A(2, 1), B(5, 1) and C(5, 5).
   b) (i) Draw the image of triangle ABC under the transformation represented by the matrix $\begin{pmatrix} 0 & 1 \\ 1 & 0 \end{pmatrix}$ and label it $A_1B_1C_1$.
      (ii) Describe this single transformation.
   c) (i) Draw the image of triangle ABC under reflection in the line $y = -x$ and label it $A_2B_2C_2$.
      (ii) Find the matrix which represents this transformation.
   d) (i) Describe fully the single transformation which maps triangle ABC onto triangle $A_2B_2C_2$.
      (ii) Find the matrix which represents this transformation.

2. Answer the whole of this question on a sheet of graph paper.
   a) Draw x and y axes from -6 to $+6$, using 1 cm to represent 1 unit on each axis.

      Draw the triangle whose vertices are A(2, 2), B(5, 2) and C(5, 3).
   b) **M** is the matrix $\begin{pmatrix} 0 & -1 \\ 1 & 0 \end{pmatrix}$ which represents the transformation **T**.

      Draw accurately the image of triangle ABC under the transformation **T**, labelling it PQR.
   c) **N** is the matrix $\begin{pmatrix} 1 & 0 \\ 0 & -1 \end{pmatrix}$ which represents the transformation **U**.

      Draw accurately the image of triangle ABC under the transformation **U**, labelling it XYZ.

d) (i) Describe fully the single transformation which maps triangle PQR onto triangle XYZ.
(ii) Find the matrix which represents this transformation.

e) (i) Calculate the matrix **NM**.
(ii) This matrix represents the transformation **V**. Draw accurately the image of triangle ABC under transformation **V**, labelling it FGH.
(iii) State whether the transformation **V** is equivalent to transformation **T** followed by transformation **U** or to transformation **U** followed by transformation **T**.

Each of the questions in Exercise 24 contained several parts so you probably found them time-consuming. However, I hope that you obtained the correct answers. Check your answers at the end of this module.

Summary

In this unit on matrices you started by developing an understanding of what a matrix is and how to work with them. You should know about these special types of matrices:

- square matrix
- diagonal matrix
- unit matrix
- zero matrix
- equal matrices.

To add or subtract matrices all you do is add or subtract the corresponding elements in each matrix, provided the matrices are of the same order. Multiplication is trickier and matrices can only be multiplied if the number of columns of the first matrix equals the number of rows of the second matrix.

After mastering the technique of multiplication you learnt:

- the determinant $|\mathbf{A}|$ of a matrix $\mathbf{A} = \begin{pmatrix} a & b \\ c & d \end{pmatrix}$ is $ad - bc$
- the inverse $\mathbf{A}^{-1}$ of a matrix $\mathbf{A} = \begin{pmatrix} a & b \\ c & d \end{pmatrix}$ is $\frac{1}{|\mathbf{A}|}\begin{pmatrix} d & -b \\ -c & a \end{pmatrix}$.

In the final section of this unit I explained to you how to use matrices for geometrical transformations. You do not need to remember each transformation matrix. What you do need to know is how each transformation affects the position vectors (1, 0) and (0, 1) and how to combine the results to form the transformation vector. In each of the following transformations you obtain the new position vector by pre-multiplying the existing position vector by the transformation matrix:

- reflection in the x-axis
- reflection in the y-axis
- reflection in the line $y = x$

- reflection in the line $y = -x$
- rotation about the origin through a number of degrees in a clockwise or anticlockwise direction
- a one-way stretch parallel to the x-axis, with scale factor k and the y-axis invariant
- a one-way stretch parallel to the y-axis, with scale factor k and the x-axis invariant
- enlargement with scale factor k and centre of enlargement (0, 0)
- shear in x-direction, x-axis invariant and scale factor k
- shear in y-direction, y-axis invariant and scale factor k.

The one exception to this method of obtaining a transformation is in translation. In this case every point moves. To obtain the new position of each point you add the vector of translation to the position vector of the point.

Finally, you learnt that to perform a transformation **A** and then a transformation **B** you write this as **BA** and you can find the combined transformation by determining **BA**.

The final exercise in this module consists of four questions on matrices taken from IGCSE examination papers. If you have followed the work in the module, you should be confident that you can cope with these questions.

Check your progress

1. $A = (1\ \ 2)$, $B = \begin{pmatrix} -3 \\ 4 \end{pmatrix}$ and $C = \begin{pmatrix} -2 & 5 \\ -3 & 6 \end{pmatrix}$.
   a) Which one of the following matrix calculations is possible?
      (i) $A + B$ (ii) AC (iii) BC
   b) Calculate AB.
   c) Find C^{-1}, the inverse of C.

2. a) Solve for x and y $\begin{pmatrix} 3 & 2 \\ -1 & 6 \end{pmatrix}\begin{pmatrix} -3 \\ 2 \end{pmatrix} = \begin{pmatrix} x \\ y \end{pmatrix}$.
   b) Find the inverse of the matrix $\begin{pmatrix} 2 & -1 \\ 4 & 3 \end{pmatrix}$.
   c) Solve for t and u $\begin{pmatrix} 3t & u \\ -t & 3u \end{pmatrix}\begin{pmatrix} 1 \\ 2 \end{pmatrix} = \begin{pmatrix} 10 \\ -10 \end{pmatrix}$.

3. Answer the whole of this question on a sheet of graph paper.
   a) Draw x and y axes from -6 to $+6$ using a scale of 1 cm to represent 1 unit on each axis.
      Draw triangle ABC with A(1, 1), B(4, 1) and C(4, 2).
   b) (i) Draw the image of triangle ABC when it is rotated 90° anticlockwise about the origin. Label this image $A_1B_1C_1$.

(ii) Triangle $A_1B_1C_1$ is translated by the vector $\begin{pmatrix} 3 \\ 1 \end{pmatrix}$. Draw and label this image $A_2B_2C_2$.

(iii) Describe fully the single transformation which maps triangle ABC onto triangle $A_2B_2C_2$.

c) (i) Draw the image of triangle ABC under the transformation represented by the matrix $\begin{pmatrix} 1 & 0 \\ 0 & -2 \end{pmatrix}$. Label this image $A_3B_3C_3$.

(ii) Describe fully the single transformation which maps triangle ABC onto triangle $A_3B_3C_3$.

4. Answer the whole of this question on a sheet of graph paper.

a) Draw axes from −6 to +6, using a scale of 1 cm to represent 1 unit on each axis.

(i) Plot the points A(5, 0), B(1, 3) and C(−1, 2) and draw triangle ABC.

(ii) Plot the points A′(3, 4), B′(3, −1) and C′(1, −2) and draw triangle A′B′C′.

b) (i) Draw and label the line l in which the triangle A′B′C′ is a reflection of triangle ABC.

(ii) Write down the equation of the line l.

(iii) Find the values of p, q, r and s such that:

$$\begin{pmatrix} p & q \\ r & s \end{pmatrix} \overset{\text{A}\quad\text{B}\quad\text{C}}{\begin{pmatrix} 5 & 1 & -1 \\ 0 & 3 & 2 \end{pmatrix}} = \overset{\text{A}'\quad\text{B}'\quad\text{C}'}{\begin{pmatrix} 3 & 3 & 1 \\ 4 & -1 & -2 \end{pmatrix}}$$

(iv) What transformation does the matrix $\begin{pmatrix} p & q \\ r & s \end{pmatrix}$ represent?

c) Reflect triangle A′B′C′ in the y-axis. Label the new triangle A″B″C″.

d) If triangle ABC is rotated about the origin, it will map onto triangle A″B″C″.

What is the angle of rotation?

If you have answered these questions correctly, you should feel pleased. Check your answers at the end of this module.

Solutions

EXERCISE 1

1. a)

Number of people	Tally marks	Number of houses
1	||	2
2	~~||||~~ ~~||||~~	10
3	~~||||~~ ||	7
4	~~||||~~ ~~||||~~ |||	13
5	~~||||~~	5
6	|||	3

b)

Number of houses

0 1 2 3 4 5 6 7 8 9 10 11 12 13

1 2 3 4 5 6

Number of people

1. c) Add up the 40 numbers recorded or use the answer to part a) as follows:

Total number of people

$=(1\times 2)+(2\times 10)+(3\times 7)+(4\times 13)+(5\times 5)+(6\times 3)$
$= 2 + 20 + 21 + 52 + 25 + 18$
$= 138$

2. a) Boys with shoe size 12 = 6
b) Total number of boys
$= 7 + 3 + 1 + 2 + 0 + 2 + 1 + 6 = 22$
c) It is not the shape expected.
You would expect there to be more boys with shoe sizes in the middle of the range (sizes 7, 8, 9, 10) and fewer size 5 and size 12.

EXERCISE 2

1. Total number of students = 1080.
Angle for:

Science $= \frac{495}{1080} \times 360° = 165°$

Arts $= \frac{375}{1080} \times 360° = 125°$

Medicine $= \frac{108}{1080} \times 360° = 36°$

Engineering $= \frac{54}{1080} \times 360° = 18°$

Law $= \frac{48}{1080} \times 360° = 16°$

(Check: Total = 360°)

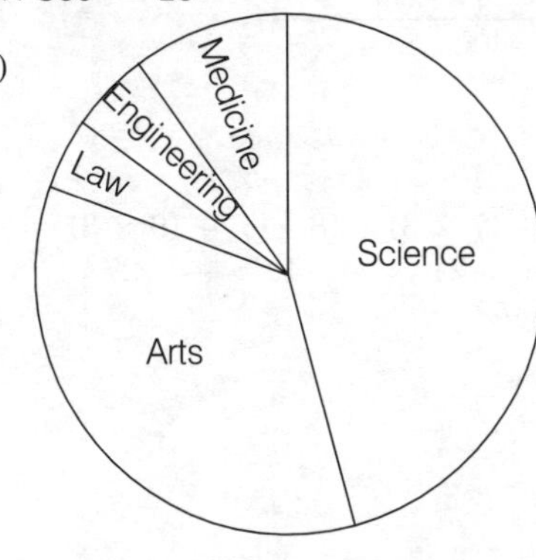

2.

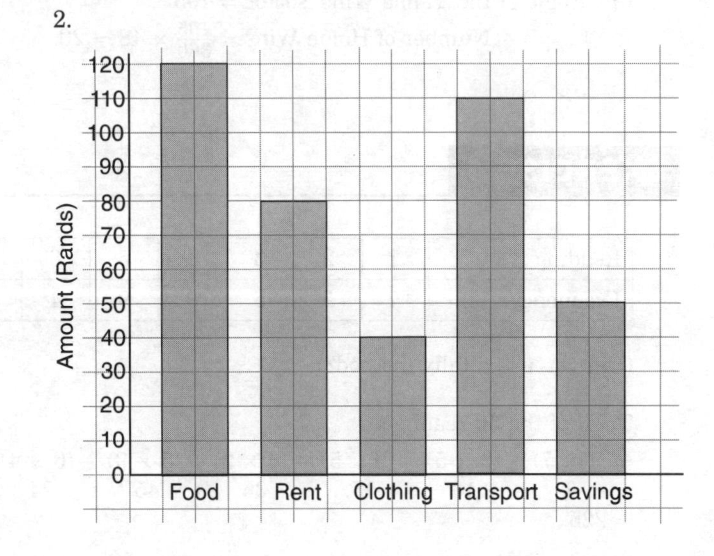

EXERCISE 2 (cont.)

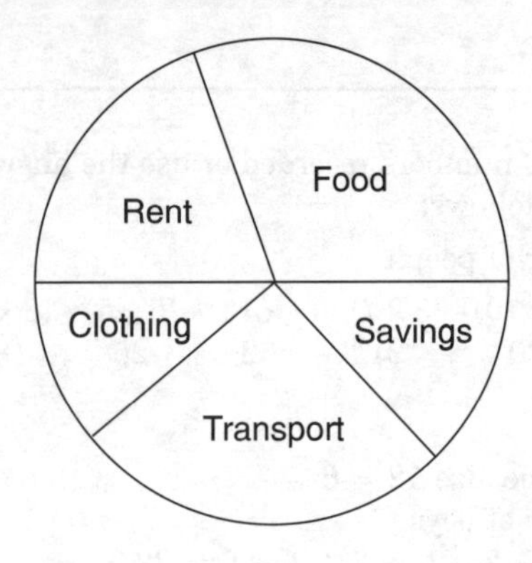

The angles in the pie chart are

Food	108°
Rent	72°
Clothing	36°
Transport	99°
Savings	45°

3. Fraction spent on wages $= \frac{150}{360} = \frac{5}{12}$

Amount spent on wages $=$ R720 000 $\times \frac{5}{12} =$ R300 000

Similarly:

Amount spent on raw materials $=$ R720 000 $\times \frac{120}{360} =$ R240 000

Amount spent on fuel $=$ R720 000 $\times \frac{40}{360} =$ R80 000

Amount spent on extras $=$ R720 000 $\times \frac{50}{360} =$ R100 000

(Check: Total $=$ R720 000)

4. a) $\frac{\text{Area of Woodland}}{\text{Total area of district}} = \frac{80}{360} = \frac{2}{9}$

b) Angle of the 'Urban' sector $= 360° - (150° + 80° + 50°) = 80°$

c) Considering the 'Urban' sector,
80° represents 50 km^2
so 360° represents $\frac{50}{80} \times 360$ km^2 which is 225 km^2

Total area of the district $=$ 225 km^2

EXERCISE 3

1. a) Fraction who said swimming $= 1 - (\frac{1}{4} + \frac{1}{8} + \frac{1}{3})$

$= 1 - \frac{6+3+8}{24}$

$= 1 - \frac{17}{24}$

$= \frac{7}{24}$

b) $x = \frac{1}{8}$ of $360° = 45°$

c) $\frac{1}{3}$ of the students chose football, so the total number of students $= 32 \times 3 = 96$

Number who said tennis $= \frac{1}{4}$ of $96 = 24$

2. a) 75°

b) Angle of the 'Home Wins' sector $= 195°$

Number of Home Wins $= \frac{195}{360} \times 48 = 26$

3. a) Belgian (The angle of the Belgian sector is 40°.)

b) (i) Fraction of the people who were French $= \frac{60}{360} = \frac{1}{6}$

(ii) $\frac{1}{6} = \frac{1}{6} \times \frac{100}{100} = 16.666..\% = 17\%$ to the nearest whole number

c) $\frac{\text{Number of Germans}}{\text{Total number of people}} = \frac{135}{360} = 0.375$

d) Number who were Dutch $= \frac{45}{360} \times 288 = 36$

4. a) Number sold in Week 2 $= 340$

Number sold in Week 3 $= 425$ (the small piece is $\frac{1}{4}$ of a loaf)

b) Week 4

EXERCISE 4

1.

Reading	1	2	3	4	5	6	7	8	9
Frequency	7	5	5	9	9	4	3	5	3

(Did you use a tally method?)

Total of the 50 readings

$= (1 \times 7) + (2 \times 5) + (3 \times 5) + (4 \times 9) + (5 \times 9) + (6 \times 4) + (7 \times 3) + (8 \times 5) + (9 \times 3)$

$= 7 + 10 + 15 + 36 + 45 + 24 + 21 + 40 + 27$

$= 225$

Mean $= \frac{225}{50} = 4.5$

EXERCISE 4 (cont.)

2. a) In numerical order: 1, 2, 3, (5,) 6, 7, 8 — Median = 5

   b) In numerical order: 43, 60, 70, (75, 85,) 95, 99, 100 — Median $= \frac{75+85}{2} = 80$

   c) In numerical order: 1, 2, (3, 4,) 5, 6 — Median $= \frac{3+4}{2} = 3.5$

   d) In numerical order: 20, 21, (22, 25,) 28, 31 — Median $= \frac{22+25}{2} = 23.5$

3. a) The distribution is

Reading	1	2	3	4	5	6	7	8	9
Frequency	1	1	0	2	6	1	1	0	1

   Mode = 5

   b) There is no mode.

   c) There are two modes: 2 and 5.

4. a) Mode = 0

   b) Median $= \frac{1}{2}$ (30th reading + 31st reading)

   30th reading in ascending order = 1.
   31st reading in ascending order = 1.
   Hence, median = 1.

   c) Total number of letters $= (0 \times 28) + (1 \times 21) + (2 \times 6) + (3 \times 3) + (4 \times 1) + (5 \times 1)$

   $= 0 + 21 + 12 + 9 + 4 + 5$

   $= 51$

   Mean $= \frac{51}{60} = 0.85$

EXERCISE 5

1.

Mass (m grams)	$60 \le m < 63$	$63 \le m < 64$	$64 \le m < 65$	$65 \le m < 66$	$66 \le m < 68$	$68 \le m < 72$
Frequency	9	12	15	17	10	8
Class width (w)	3	1	1	1	2	4
Height of rectangle (f/w)	3	12	15	17	5	2

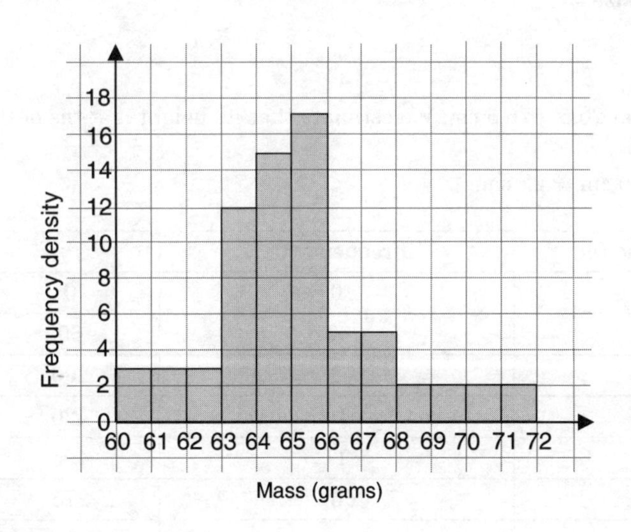

EXERCISE 5 (cont.)

2. a) The rectangle for the interval $14\frac{1}{2} - 19\frac{1}{2}$ contains 16 squares and it represents a frequency of 8.
The rectangle for the interval $19\frac{1}{2} - 24\frac{1}{2}$ contains 14 squares and it represents a frequency of 7.
So, each square represents a frequency of $\frac{1}{2}$.
The rectangle for the interval $4\frac{1}{2} - 14\frac{1}{2}$ must represent a frequency of 4 and so it contains 8 squares. Its base is 4 squares wide and so its height must be 2 squares.

Similarly, the rectangle for the interval $24\frac{1}{2} - 39\frac{1}{2}$, which must represent a frequency of 6, contains 12 squares.

Its base is 6 squares wide, so its height is 2 squares.

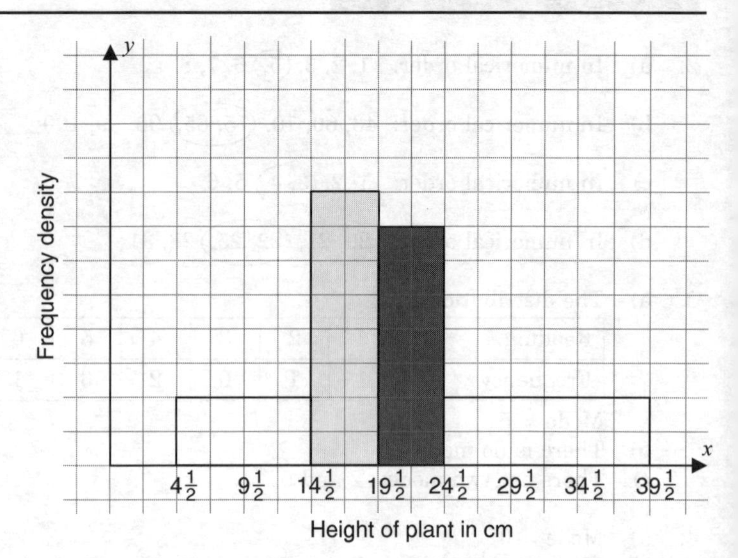

b)

Height in cm	Mid-interval value (x)	Frequency (f)	fx
5–14	9.5	4	38
15–19	17	8	136
20–24	22	7	154
25–39	32	6	192
		25	520

Mean height of the plants $= \frac{520}{25}$ cm $= 20.8$ cm

EXERCISE 6

1. a)

Height (h) in cm	$h < 5.5$	$h < 15.5$	$h < 20.5$	$h < 25.5$	$h < 40.5$
Cumulative frequency	0	3	10	20	25

b) The median is the height of the 13th plant in order of size.
There are 10 plants with height less than 20.5 cm
and 20 plants with height less than 25.5 cm
The median height is in the interval 20.5 cm to 25.5 cm.

c) The 13th plant is the 3rd plant (out of 10) in the interval 20.5 – 25.5 cm. We estimate that its height is $\frac{3}{10}$ths of the length of the interval from the lower end of the interval.
Estimate of the median plant height $= (20.5 + (\frac{3}{10} \times 5))$ cm $= 22$ cm.

2. a)

Amount spent (R)	Mid-interval value (m)	Frequency (f)	fx
0–10	5	0	0
10–20	15	4	60
20–30	25	8	200
30–40	35	12	420
40–50	45	11	495
50–60	55	5	275
		40	1450

Mean $= \frac{\text{R}1450}{40} = \text{R}36.25$

b) $p = 0 + 4 + 8 = 12$
$q = 0 + 4 + 8 + 12 = 24$
$r = 0 + 4 + 8 + 12 + 11 = 35$

EXERCISE 6 (cont.)

2. c)

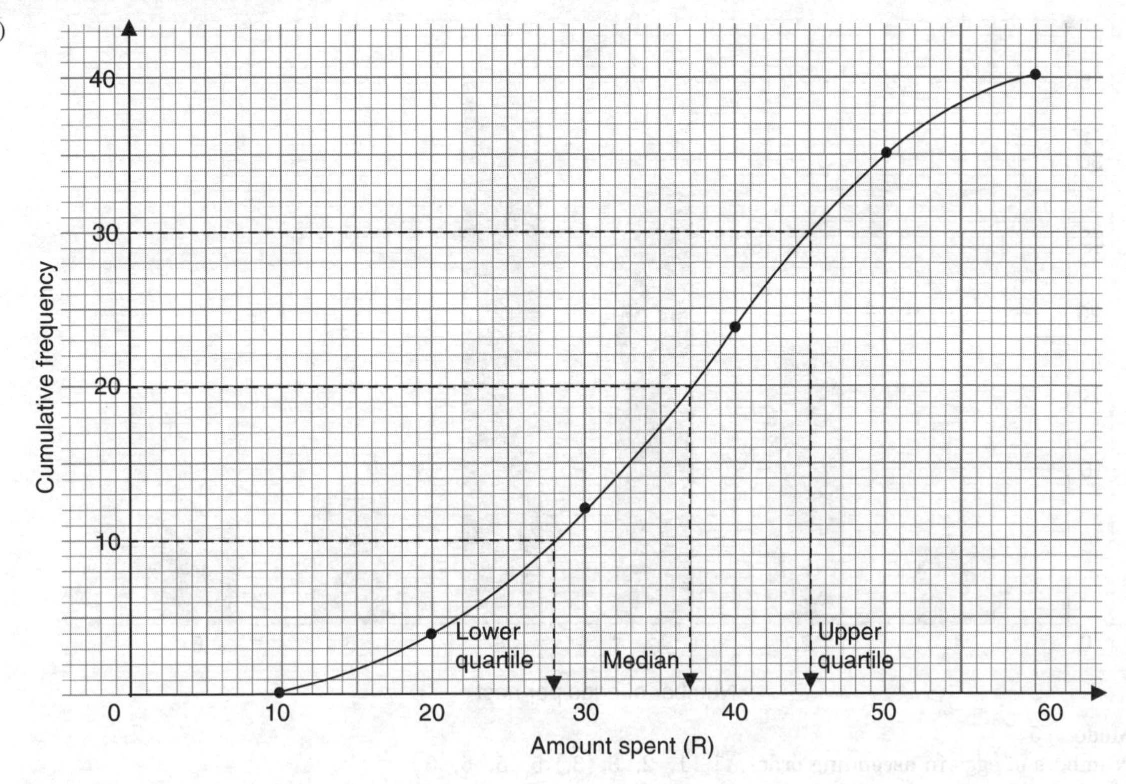

d) (i) Median = R37

(ii) Upper quartile = R45
Lower quartile = R28
Inter-quartile range = R45 − R28 = R17

Check your progress 1

1. a) $x^\circ + 144^\circ + 90^\circ = 360^\circ$
Hence, $x = 360 - 234 = 126$

b) Number who chose rabbits $= \frac{90}{360} \times 200 = 50$.

c) Fraction who chose dogs $= \frac{144}{360} = \frac{2}{5}$.

Percentage who chose dogs $= \frac{2}{5} \times \frac{100}{100} = 40\%$.

2. a) 2

b) 2 teams scored 0 goals.
4 teams scored 1 goal each.
1 team scored 2 goals.
3 teams scored 3 goals each.
2 teams scored 5 goals each.
Total number of teams = 2 + 4 + 1 + 3 + 2 = 12

c) Total number of goals
$= (0 \times 2) + (1 \times 4) + (2 \times 1) + (3 \times 3) + (5 \times 2)$
$= 0 + 4 + 2 + 9 + 10$
$= 25$

Check your progress 1 (cont.)

3. a)

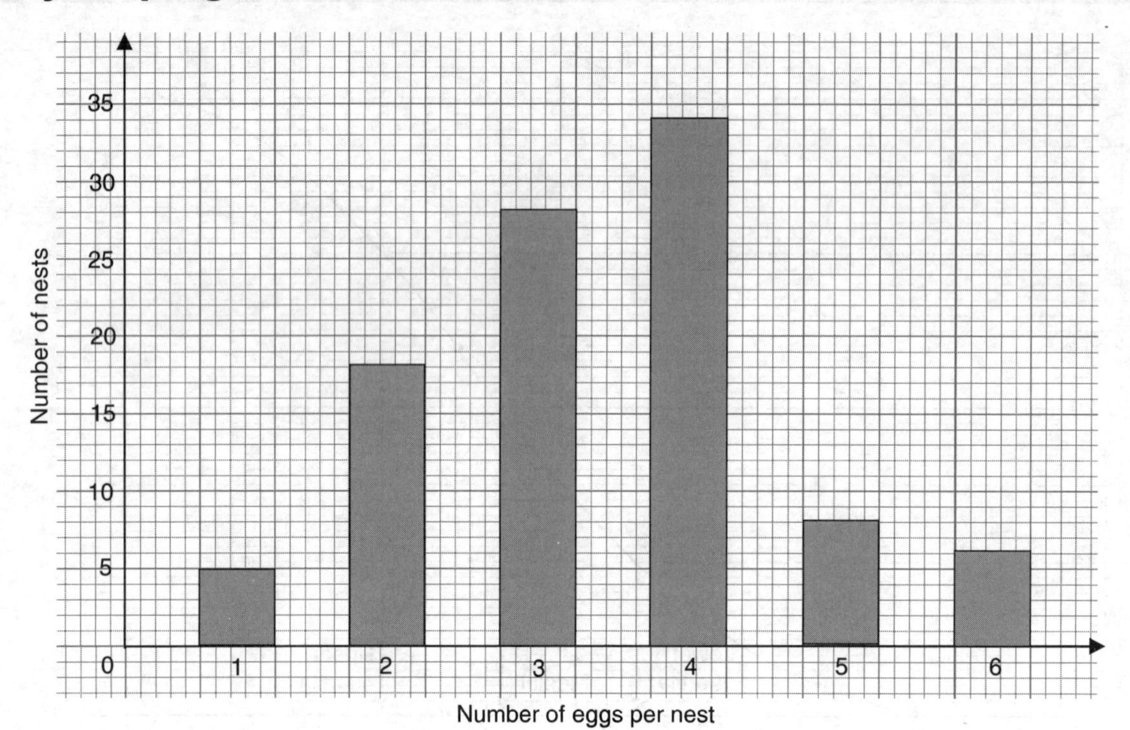

b) (i) Mode = 5

(ii) Numbers of eggs in ascending order: 1, 1, 2, 3, (3,) 5, 5, 5, 6

Median = 3

(iii) Total number of eggs = 31

Number of nests = 9

Mean = $\frac{31}{9}$ = 3.4444. . . = 3.4 to 1 decimal place.

4. Median = 5 indicates that 5 is the middle number (of the three numbers).
Mean = 6 indicates that the sum of the three numbers is 18.
Hence, the first and third numbers must add up to 13.
The first number must be less than 5 and the third number must be more than 5.
The positive integers could be 4, 5, 9 or 3, 5, 10 or 2, 5, 11 or 1, 5, 12.
(The question only requires you to give one of these answers.)

5. a)

Mass (M) in grams	Number of fish	Classification
$M < 300$	$1 + 2 = 3$	Small
$300 \leq M < 400$	$5 + 4 = 9$	Medium
$M \geq 400$	$3 + 3 = 6$	Large

b) Total number of fish = 18

Angle for:

Small fish = $\frac{3}{18} \times 360° = 60°$

Medium fish = $\frac{9}{18} \times 360° = 180°$

Large fish = $\frac{6}{18} \times 360° = 120°$

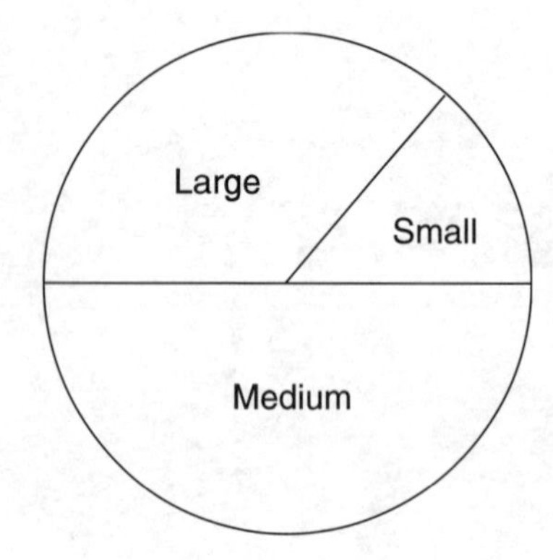

Check your progress 1 (cont.)

6. a) Mode = 1
   b) There are 40 families with 0 or 1 child.
   There are 67 families with 0 or 1 or 2 children.
   The 50th family has 2 children and the 51st family has 2 children.
   Hence, median = 2.
   c) Total number of children $= (0 \times 4) + (1 \times 36) + (2 \times 27) + (3 \times 21) + (4 \times 5) + (5 \times 4) + (6 \times 2) + (7 \times 1)$
   $= 0 + 36 + 54 + 63 + 20 + 20 + 12 + 7$
   $= 212$
   Number of families = 100
   Mean number of children $= \frac{212}{100} = 2.12$

7. a)

Time (minutes)	Mid-interval value (x)	Frequency (f)	fx
0–1	0.5	12	6
1–2	1.5	14	21
2–4	3	20	60
4–6	5	14	70
6–8	7	12	84
8–10	9	18	162
10–15	12.5	10	125
		100	528

(i) Mean $= \frac{528}{100}$ minutes = 5.28 minutes

(ii) 0.28 minutes = 0.28×60 seconds = 16.8 seconds

Mean = 5 minutes 16.8 seconds
= 5 minutes 17 seconds to the nearest second.

b)

Time (t minutes)	Cumulative frequency
0	0
≤ 1	12
≤ 2	26
≤ 4	46
≤ 6	60
≤ 8	72
≤ 10	90
≤ 15	100

Check your progress 1 (cont.)

7. c)

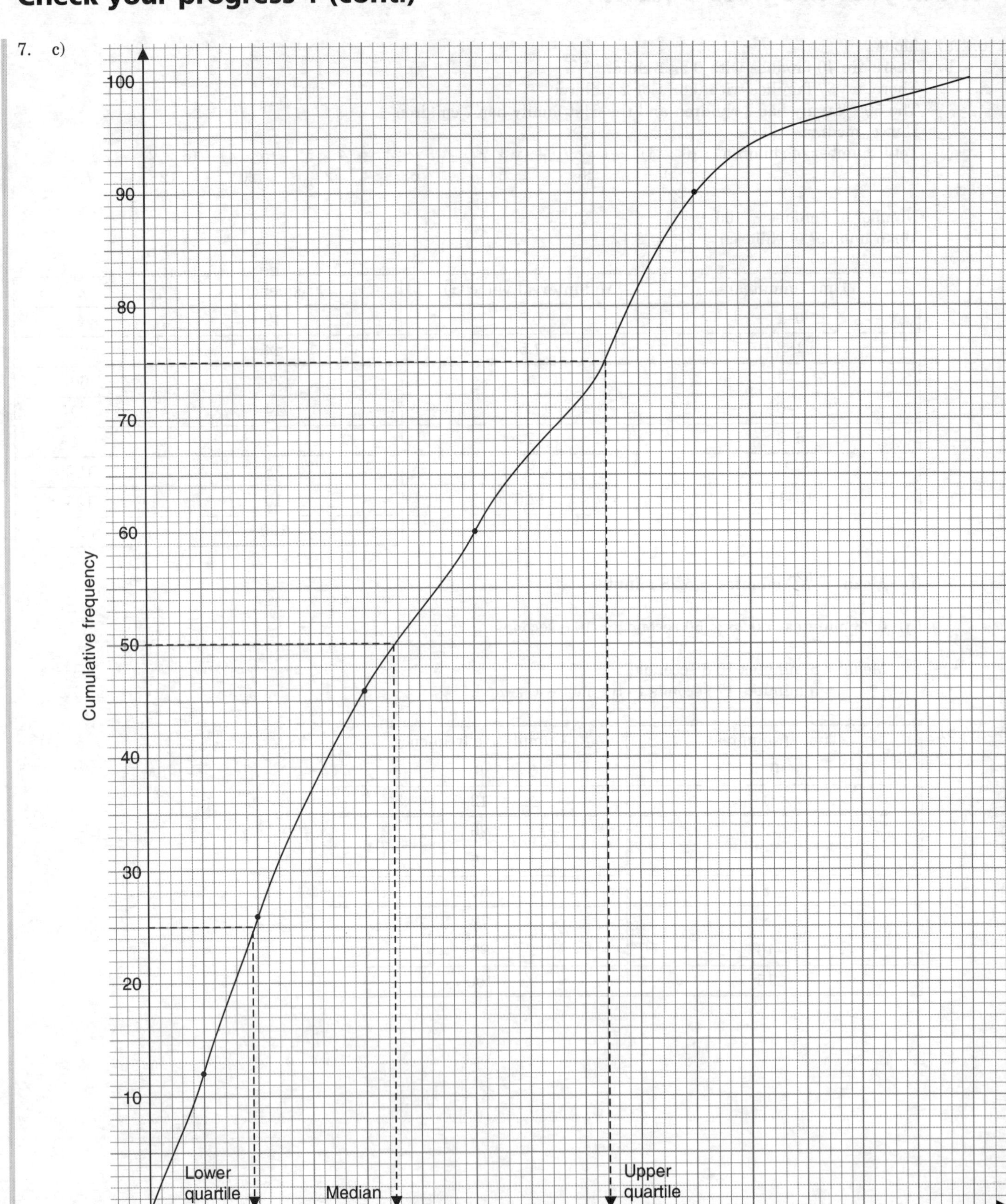

d) (i) Median = 4.5 minutes

(ii) Upper quartile = 8.4 minutes

Lower quartile = 1.9 minutes

Inter-quartile range = (8.4 − 1.9) minutes

= 6.5 minutes

EXERCISE 7

(In most cases, probabilities are given as fractions. You could give them as decimals if they are exact.)

1. There are four kings among the 52 cards.
   Probability of drawing a king $= \frac{4}{52} = \frac{1}{13}$.

2. There are six possible equally likely scores (1, 2, 3, 4, 5, 6) and three of them are prime numbers (2, 3, 5).
   Probability of turning up a prime number $= \frac{3}{6} = \frac{1}{2}$.

3. Total number of balls = 10.
   a) Number of red balls = 2.
   P(red ball) $= \frac{2}{10} = \frac{1}{5}$.
   b) Number of balls which are red or white = 5.
   P(red ball or white ball) $= \frac{5}{10} = \frac{1}{2}$.
   c) If the ball is neither red nor black, it must be white.
   Number of white balls = 3.
   P(ball neither red nor black) $= \frac{3}{10}$.

4. a) P(blue ball) $= \frac{5}{20} = \frac{1}{4}$.
   b) $\frac{1}{5}$ of the balls in the bag are red.
   Number of red balls $= 20 \times \frac{1}{5} = 4$.

5. a)

Number of matches	39	40	41	42	43	44	45
Frequency	5	3	4	3	4	0	1

   b) Number of boxes containing more than 40 matches
   $= 4 + 3 + 4 + 0 + 1$
   $= 12$
   P(box contains more than 40 matches) $= \frac{12}{20} = \frac{3}{5}$.

6. a) In every 50 seconds, the lights are green for 20 seconds.
   P(lights are green) $= \frac{20}{50} = \frac{2}{5}$.
   b) In every 50 seconds, the lights are red for 25 seconds.
   P(lights are red) $= \frac{25}{50} = \frac{1}{2}$.

EXERCISE 8

1. The 'favourable' outcomes are 1, 3, 5 and 6 (out of six possible, equally likely outcomes).
   P(dice shows 6 or an odd number) $= \frac{4}{6} = \frac{2}{3}$.

   Alternative method

   P(dice shows 6) $= \frac{1}{6}$ and P(dice shows an odd number) $= \frac{3}{6}$.
   These events are mutually exclusive, so
   P(dice shows 6 or an odd number) $= \frac{1}{6} + \frac{3}{6} = \frac{4}{6} = \frac{2}{3}$.

2. In a pack of 52 cards, there are four 4s and four kings.
   P(card is a 4 or a king) $= \frac{8}{52} = \frac{2}{13}$.

3. P(red straw) $= \frac{30}{60} = \frac{1}{2}$ and P(green straw) $= \frac{10}{60}$.
   P(red straw or green straw) $= \frac{1}{2} + \frac{1}{6} = \frac{3+1}{6} = \frac{4}{6} = \frac{2}{3}$.

4. a) P(red or blue) = P(red) + P(blue) $= \frac{1}{3} + \frac{1}{5}$
   $= \frac{5+3}{15} = \frac{8}{15}$.
   b) P(white) = P(not red or blue) $= 1 - \frac{8}{15} = \frac{7}{15}$.

5. If the coins do not show three heads, they must show at least one tail. P(at least one tail)
   = P(coins do not show three heads)
   = 1 – P(coins show three heads)
   = 1 – 0.125
   = 0.875

EXERCISE 9

1. For each toss, P(head) $= \frac{1}{2}$ and P(tail) $= \frac{1}{2}$.

   For two tosses, P(head, head) $= \frac{1}{2} \times \frac{1}{2} = \frac{1}{4}$ and

   P(tail, tail) $= \frac{1}{2} \times \frac{1}{2} = \frac{1}{4}$.

   Probability of two tosses giving same result
   = P(head, head) + P(tail, tail)

   $= \frac{1}{4} + \frac{1}{4} = \frac{1}{2}$

   Note: There are four possible outcomes of two tosses: head, head; head, tail; tail, head; tail, tail, and these are equally likely.
   Hence, P(head, head or tail, tail) $= \frac{2}{4} = \frac{1}{2}$.

2. P(first marble is blue) $= \frac{8}{10}$ are P(first marble is red) $= \frac{2}{10}$.

   If the first marble is blue, there are 9 marbles left, 7 are blue and 2 red. In this case, P(second marble is blue) $= \frac{7}{9}$ and P(second marble is red) $= \frac{2}{9}$.

   A similar calculation can be done for the case when the first marble is red.

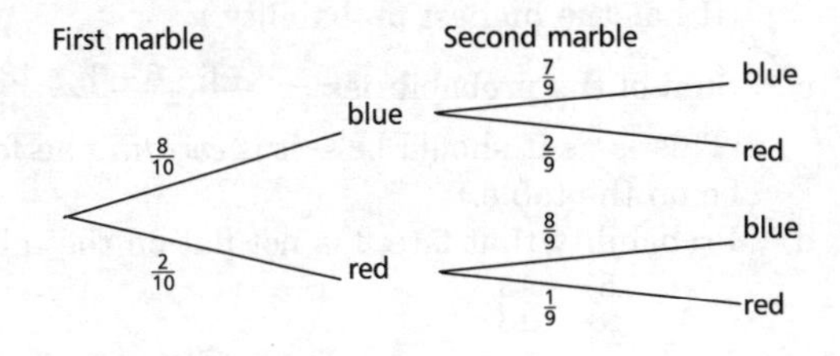

EXERCISE 9 (cont.)

2. a) P(2 red marbles) = P(red, red) $= \frac{2}{10} \times \frac{1}{9} = \frac{1}{45}$.

   b) P(1 red marble and 1 blue marble)
   = P(red, blue or blue, red)
   = P(red, blue) + P(blue, red)
   $= (\frac{2}{10} \times \frac{8}{9}) + (\frac{8}{10} \times \frac{2}{9})$
   $= \frac{8}{45} + \frac{8}{45}$
   $= \frac{16}{45}$

   c) P(2 blue marbles) = P(blue, blue) $= \frac{8}{10} \times \frac{7}{9} = \frac{28}{45}$.

   Check: These three answers add up to 1 because 2 reds, 1 red and 1 blue, 2 blues are the only possible outcomes.

3.

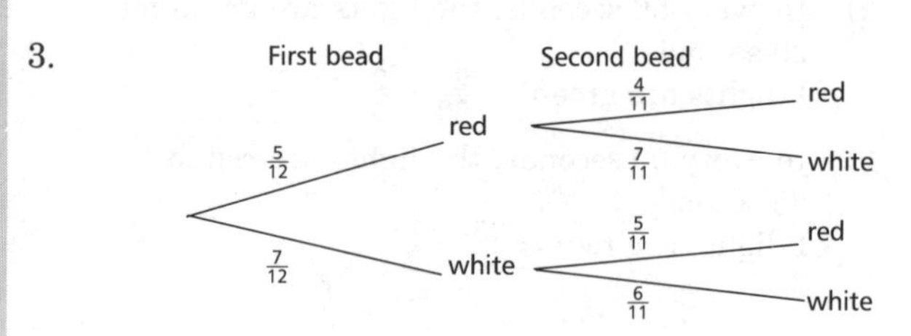

   a) P(2 red beads) $= \frac{5}{12} \times \frac{4}{11} = \frac{5}{33}$.

   b) P(2 white beads) $= \frac{7}{12} \times \frac{6}{11} = \frac{7}{22}$.

4. a) (i)

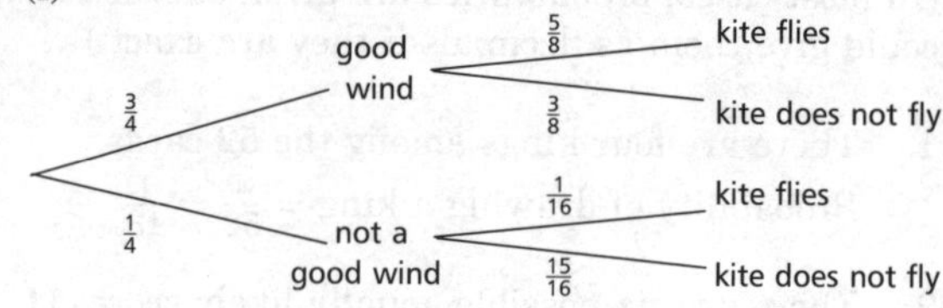

   (ii) P(good wind and kite flies) $= \frac{3}{4} \times \frac{5}{8} = \frac{15}{32}$.

   (iii) P(good wind and kite does not fly)
   $= \frac{3}{4} \times \frac{3}{8} = \frac{9}{32}$

   P(not a good wind and kite does not fly)
   $= \frac{1}{4} \times \frac{15}{16} = \frac{15}{64}$

   P(kite does not fly whatever the wind)
   $= \frac{9}{32} + \frac{15}{64} = \frac{33}{64}$

   b) P(kite flies whatever the wind)
   $= 1 - \frac{33}{64}$ or $(\frac{3}{4} \times \frac{5}{8}) + (\frac{1}{4} \times \frac{1}{16}) = \frac{31}{64}$

   P(kite sticks in a tree whatever the wind)
   $= \frac{31}{64} \times \frac{1}{2} = \frac{31}{128}$

Check your progress 2

1. a) $x° + 90° + 120° = 360°$ so $x = 360 - 210 = 150$

   b) Number who walk to school $= \frac{90}{360} \times 240 = 60$.

   c) Fraction who travel by bus $= \frac{120}{360} = \frac{1}{3}$.

   Probability that the pupil (chosen at random) travels by bus $= \frac{1}{3}$.

2. a) red, black, black; black, red, black; black, black, red.

   b) In two cases out of three the two black cards are next to one another. P(the two black cards are next to one another) $= \frac{2}{3}$.

3. a)

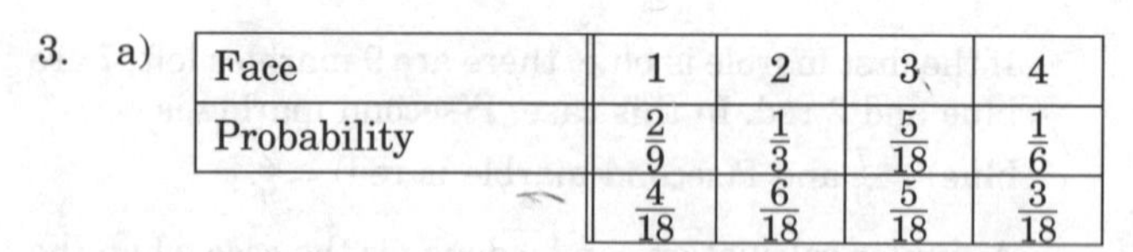

Face	1	2	3	4
Probability	$\frac{2}{9}$	$\frac{1}{3}$	$\frac{5}{18}$	$\frac{1}{6}$
	$\frac{4}{18}$	$\frac{6}{18}$	$\frac{5}{18}$	$\frac{3}{18}$

   b) Face 2 is the most likely to finish flat on the table. (It has the highest probability.)

   c) Sum of the probabilities $= \frac{4+6+5+3}{18} = \frac{18}{18} = 1$.
   (This is as it should be – it is *certain* one face will be on the table.)

   d) Probability that face 3 is not flat on the table
   $= 1 - \frac{5}{18} = \frac{13}{18}$

4. a) First die shows 6 and the second die shows 2.

   b)

Second die: 1 2 3 4 5 6
First die: 1 2 3 4 5 6
P

   c) There are 36 possible outcomes and 5 of them give a total of 8.

   (i) P(total score of 8) $= \frac{5}{36}$.

   (ii) In 6 of the 36 outcomes, the two dice show the same score (1 and 1, 2 and 2, 3 and 3, 4 and 4, 5 and 5, 6 and 6).
   P(dice show same score as each other)
   $= \frac{6}{36} = \frac{1}{6}$

Check your progress 2 (cont.)

5. a) P(Jane will not pass in English) $= 1 - \frac{3}{4} = \frac{1}{4}$.
   b) P(Jane will pass in both subjects)
   = P(Jane will pass in English) × P(Jane will pass in maths)
   $= \frac{3}{4} \times \frac{4}{5} = \frac{3}{5}$

6. a) (i) Number of girls who speak only one language = 50.
   (ii) Number of boys in the school = 35 + 25 = 60.
   b) (i) Number of students who speak more than one language = 40 + 35 = 75.
   P(student speaks more than one language
   $= \frac{75}{150} = \frac{1}{2}$.
   (ii) Number of girls in the school = 50 + 40 = 90.
   Number of girls who speak more than one language = 40.
   P(girl speaks more than one language)
   $= \frac{40}{90} = \frac{4}{9}$.
   (iii) Number of students who speak more than one language = 40 + 35 = 75.
   Of these, 40 are girls.
   P(student who speaks more than one language is a girl) $= \frac{40}{75} = \frac{8}{15}$.
   (iv) Of the 150 students, 60 are boys.
   P(first student is a boy) $= \frac{60}{150}$

   If the first student chosen is a boy, there are 59 boys among the remaining 149 students.
   In this case P(second student is a boy) $= \frac{59}{149}$

   P(both students chosen are boys)
   $= \frac{60}{150} \times \frac{59}{149} = \frac{2}{5} \times \frac{59}{149} = \frac{118}{745}$

7. a) (i) There are 2 end chairs out of 6.
   P(Alain sits on an end chair) $= \frac{2}{6} = \frac{1}{3}$.
   (ii) When Alain is seated, Bernard has 5 chairs to choose from.
   (a) If Alain sits at one end, there is only 1 chair next to his.
   So P(Bernard sits next to Alain) $= \frac{1}{5}$.
   (b) If Alain is not sitting at an end, there are 2 chairs next to his, so P(Bernard sits next to Alain) $= \frac{2}{5}$.
   (iii)

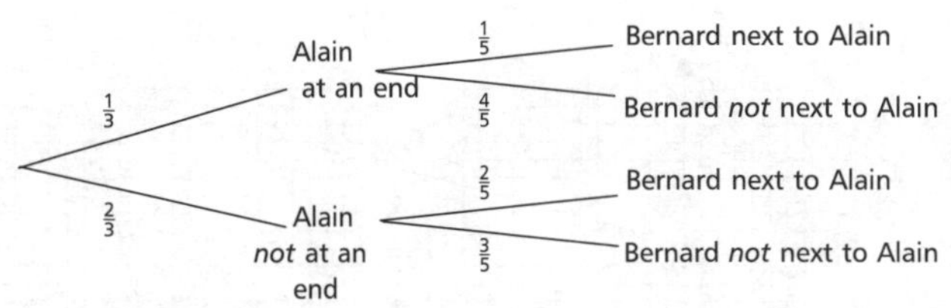

   (iv) P(Bernard sits next to Alain)
   $= (\frac{1}{3} \times \frac{1}{5}) + (\frac{2}{3} \times \frac{2}{5})$
   $= \frac{1}{15} + \frac{4}{15} = \frac{5}{15}$
   $= \frac{1}{3}$
   b) Alain could sit in any chair. Bernard has a choice of 5 chairs, of which 2 are next to Alain.
   P(Bernard sits next to Alain) $= \frac{2}{5}$.
   c) Alain could sit in any chair.
   Bernard has a choice of $(n - 1)$ chairs, of which 2 are next to Alain.
   P(Bernard sits next to Alain) $= \frac{2}{(n-1)}$
   This probability is given to be $\frac{1}{4}$.
   $\frac{2}{(n-1)} = \frac{1}{4}$ gives $8 = (n - 1)$ and so $n = 9$.

EXERCISE 10

1. a)

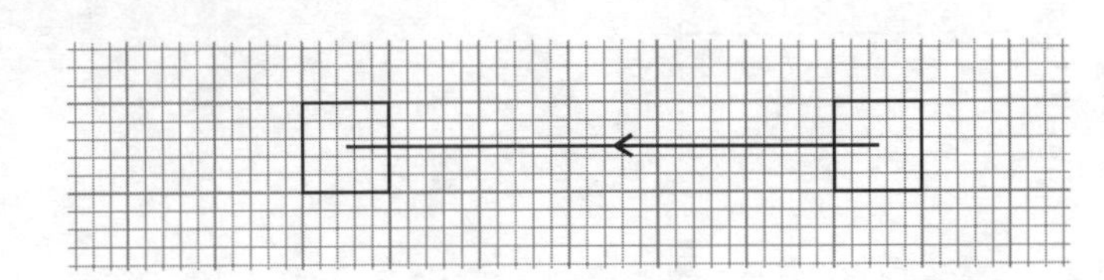

   b)

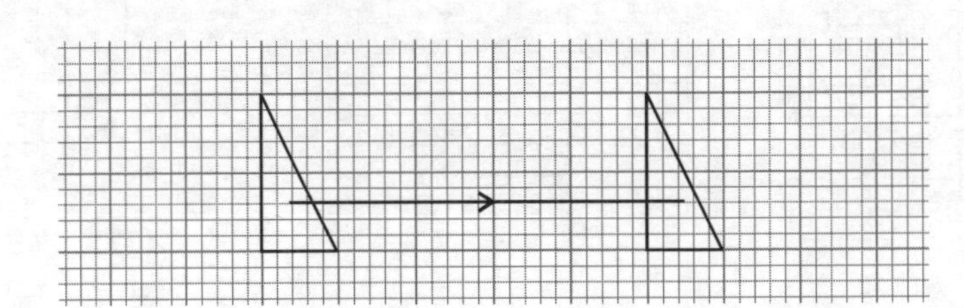

EXERCISE 10 (cont.)

2.

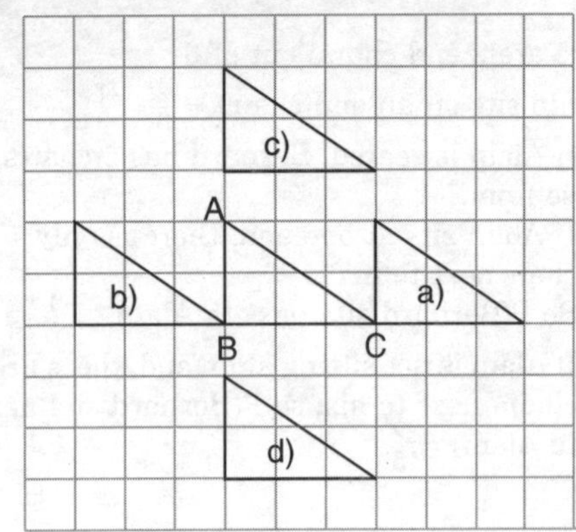

3.

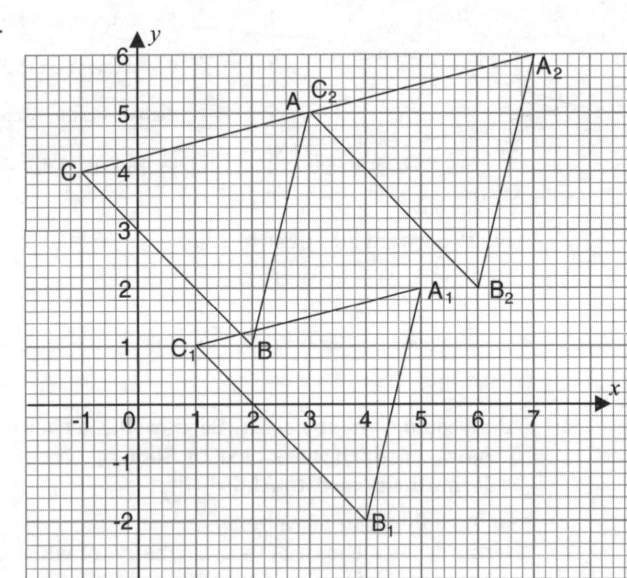

A_1 is (5, 2)

B_1 is (4, – 2)

C_1 is (1, 1)

A_2 is (7, 6)

B_2 is (6, 2)

C_2 is (3, 5), the same point as A.

EXERCISE 11

1. a)

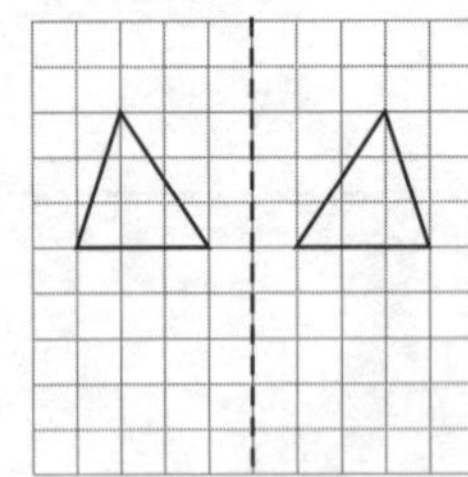

b)

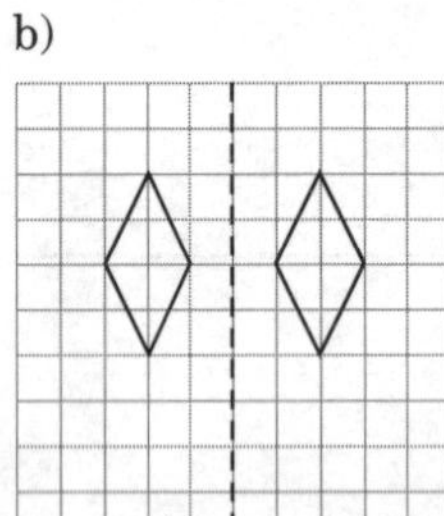

c)

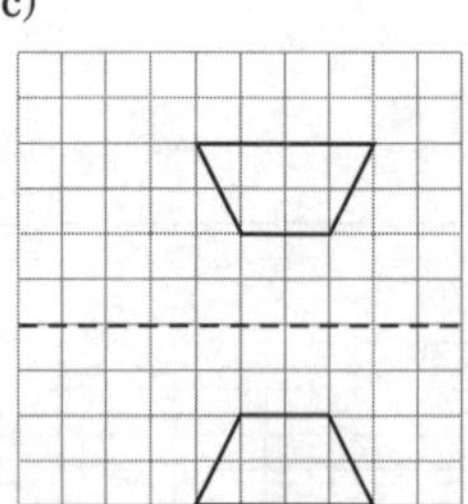

2. a)

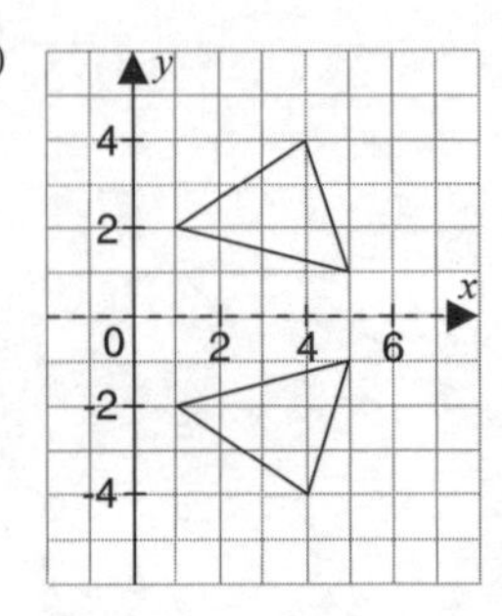

b)

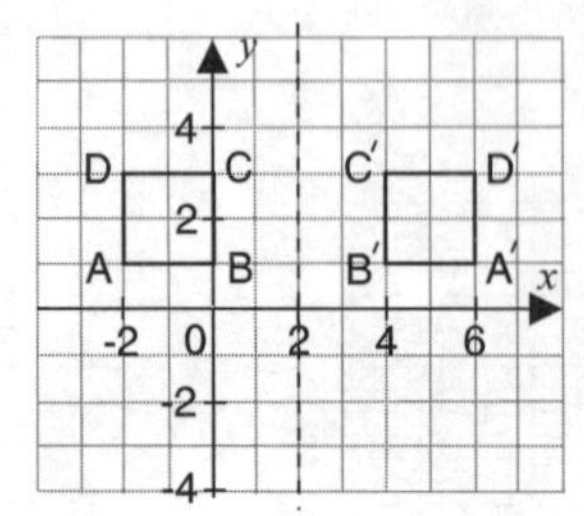

c)

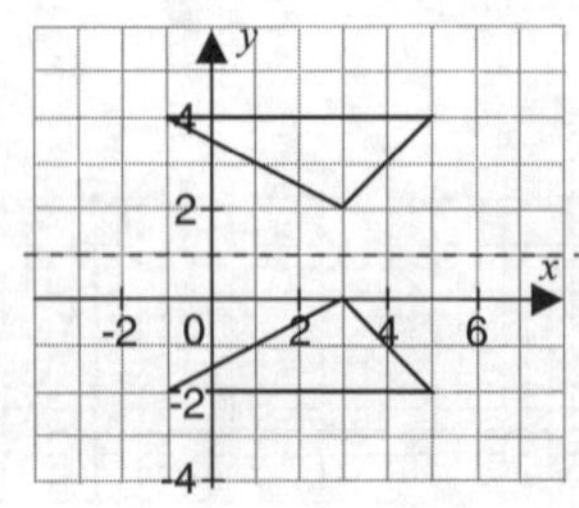

d)

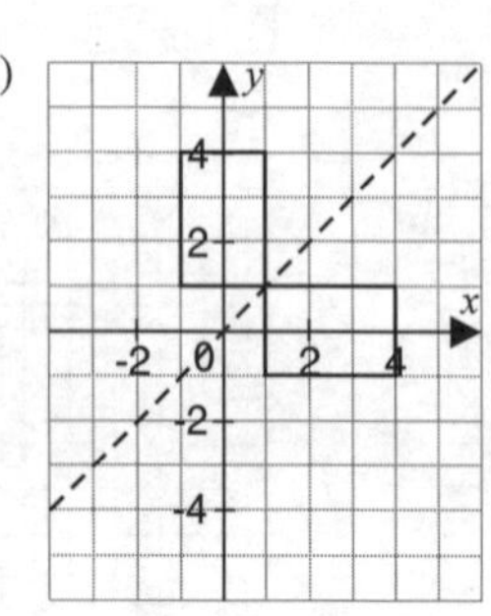

EXERCISE 12

1. a)

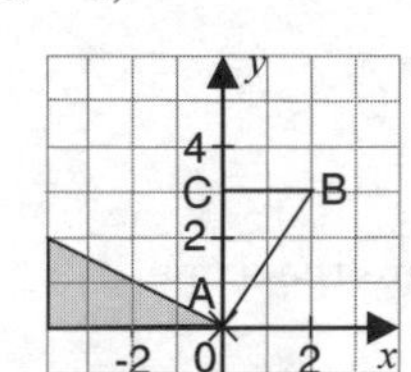

b)

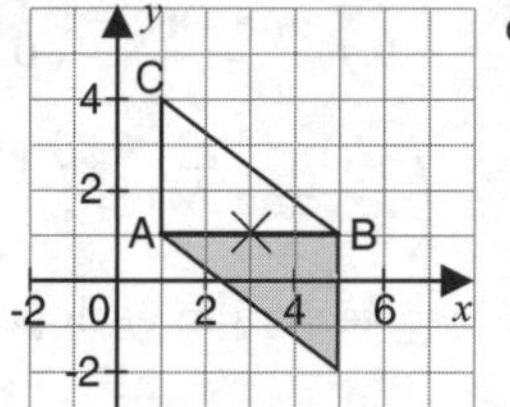

c)

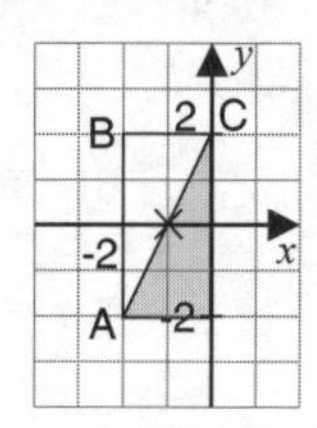

2. a) 270° anticlockwise or 90° clockwise.
(You could write this answer as +270° or −90°.)
b) 180°. (You could have 180° anticlockwise or 180° clockwise or 180° or −180°.)
c) 270° anticlockwise or 90° clockwise.
(You could have 270° or −90°.)
d) 270° anticlockwise or 90° clockwise.
(Again, you could have 270° or −90°.)

EXERCISE 13

1.

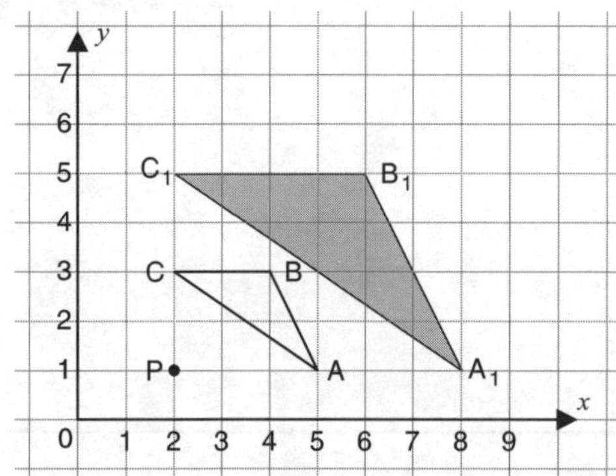

Notice that the sides of the image triangle are parallel to the corresponding sides of triangle ABC, and twice as long.

If the image is $A_1B_1C_1$, then $A_1B_1 = 2AB$, $B_1C_1 = 2BC$, $C_1A_1 = 2CA$.

2.

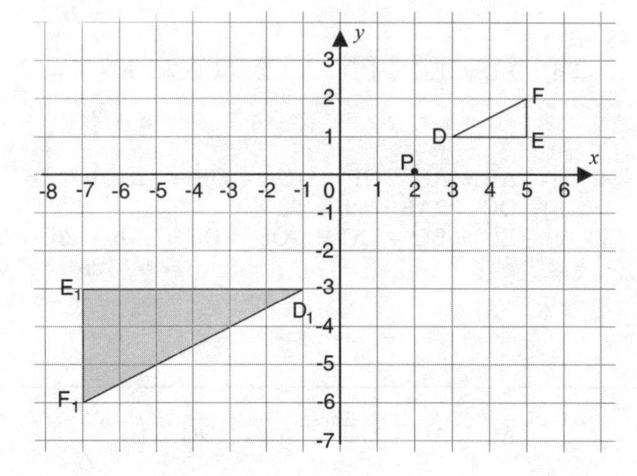

If the image is $D_1E_1F_1$, then $D_1E_1 = 3DE$, $E_1F_1 = 3EF$, $F_1D_1 = 3FD$.
Notice that because the scale factor is negative the enlargement of DEF is on the opposite side of P.

3.

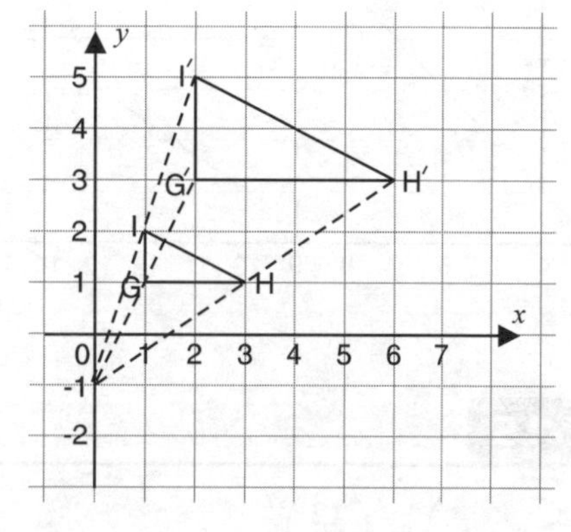

Join G′ to G, H′ to H and I′ to I and produce these lines until they meet. The point of intersection is the centre of enlargement.

Hence, the centre of enlargement is (0, – 1).

$G'H' = 2GH$, $H'I' = 2HI$, $I'G' = 2IG$ so the scale factor of the enlargement is 2 and it is positive because the triangle and its enlargement are both on the same side of the centre of enlargement.

4. The length of $P'Q' = 9$ and the length of $PQ = 6$.

Hence, the scale factor of the enlargement $= \frac{9}{6} = 1.5$.

The lines P′P, Q′Q, R′R, S′S meet at (4, 2).
Hence, the centre of the enlargement is (4, 2).

EXERCISE 14

1. $\underset{\sim}{a} = \begin{pmatrix} 4 \\ 6 \end{pmatrix}$ $\underset{\sim}{b} = \begin{pmatrix} 4 \\ 2 \end{pmatrix}$ $\underset{\sim}{c} = \begin{pmatrix} -4 \\ 2 \end{pmatrix}$ $\underset{\sim}{d} = \begin{pmatrix} -4 \\ 2 \end{pmatrix}$

   $\underset{\sim}{e} = \begin{pmatrix} 6 \\ -4 \end{pmatrix}$ $\underset{\sim}{f} = \begin{pmatrix} 0 \\ 4 \end{pmatrix}$ $\underset{\sim}{g} = \begin{pmatrix} 8 \\ 4 \end{pmatrix}$ $\underset{\sim}{h} = \begin{pmatrix} 4 \\ -2 \end{pmatrix}$

2.

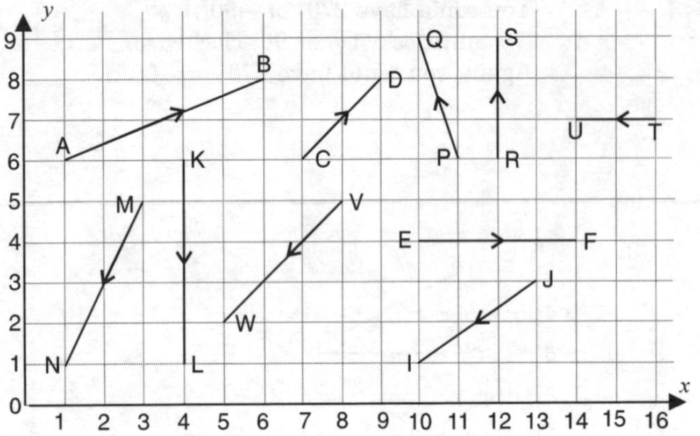

3. a) $\overrightarrow{AB} = \begin{pmatrix} 4 \\ 0 \end{pmatrix}$ and $\overrightarrow{DC} = \begin{pmatrix} 4 \\ 0 \end{pmatrix}$

   b) $\overrightarrow{BC} = \begin{pmatrix} 1 \\ 3 \end{pmatrix}$ and $\overrightarrow{AD} = \begin{pmatrix} 1 \\ 3 \end{pmatrix}$

   The two vectors in each pair are the same.

4. a) $\begin{pmatrix} 4 \\ 2 \end{pmatrix}$ b) $\begin{pmatrix} 5 \\ -1 \end{pmatrix}$ c) $\begin{pmatrix} -5 \\ 1 \end{pmatrix}$

   d) $\begin{pmatrix} 0 \\ -3 \end{pmatrix}$ e) $\begin{pmatrix} -4 \\ 3 \end{pmatrix}$ f) $\begin{pmatrix} 5 \\ 2 \end{pmatrix}$

EXERCISE 15

1. a) $3\mathbf{p} = \begin{pmatrix} 3 \times 4 \\ 3 \times (-2) \end{pmatrix} = \begin{pmatrix} 12 \\ -6 \end{pmatrix}$

   b) $\mathbf{p} + \mathbf{q} = \begin{pmatrix} 4 + (-1) \\ (-2) + (-3) \end{pmatrix} = \begin{pmatrix} 3 \\ -5 \end{pmatrix}$

2. $2\underset{\sim}{a} - \underset{\sim}{b} = 2\begin{pmatrix} 3 \\ -2 \end{pmatrix} - \begin{pmatrix} -4 \\ 3 \end{pmatrix} = \begin{pmatrix} 6 \\ -4 \end{pmatrix} - \begin{pmatrix} -4 \\ 3 \end{pmatrix}$

   $= \begin{pmatrix} 6 - (-4) \\ -4 - 3 \end{pmatrix} = \begin{pmatrix} 10 \\ -7 \end{pmatrix}$

3. a) $\overrightarrow{AC} = \overrightarrow{AB} + \overrightarrow{BC} = 2\mathbf{a} + 3\mathbf{b}$

   b) C divides AD in the ratio 2 : 1 so $\overrightarrow{CD} = \frac{1}{2}\overrightarrow{AC}$

   $= \mathbf{a} + \frac{3\mathbf{b}}{2}$

   c) C divides BE in the ratio 3 : 1 so $\overrightarrow{CE} = \frac{1}{3}\overrightarrow{BC}$

   $= \mathbf{b}.$

   d) $\overrightarrow{ED} = \overrightarrow{EC} + \overrightarrow{CD} = -\overrightarrow{CE} + \overrightarrow{CD} = -\mathbf{b} + \mathbf{a} + \frac{3\mathbf{b}}{2}$

   $= \mathbf{a} + \frac{\mathbf{b}}{2}.$

4. a) $\overrightarrow{AB} = \overrightarrow{AO} + \overrightarrow{OB} = -\overrightarrow{OA} + \overrightarrow{OB} = -\mathbf{a} + \mathbf{b}$ (or $\mathbf{b} - \mathbf{a}$)

   b) $\overrightarrow{OC} = 2\overrightarrow{AB} = 2\mathbf{b} - 2\mathbf{a}$

   c) $\overrightarrow{BC} = \overrightarrow{BO} + \overrightarrow{OC} = -\overrightarrow{OB} + \overrightarrow{OC} = -\mathbf{b} + 2\mathbf{b} - 2\mathbf{a}$

   $= \mathbf{b} - 2\mathbf{a}$

EXERCISE 16

1. $|\overrightarrow{OP}| = \sqrt{3^2 + 4^2} = \sqrt{25} = 5$

   $|\overrightarrow{OQ}| = \sqrt{(-5)^2 + (12)^2} = \sqrt{25 + 144} = \sqrt{169} = 13$

   $|\overrightarrow{OR}| = \sqrt{(-8)^2 + (-15)^2} = \sqrt{64 + 225} = \sqrt{289} = 17$

2. a) A is (4, 2), B is (−1, 3) and C is (6, −2).

   b) $\overrightarrow{AB} = \overrightarrow{AO} + \overrightarrow{OB} = -\overrightarrow{OA} + \overrightarrow{OB} = \begin{pmatrix} -4 \\ -2 \end{pmatrix} + \begin{pmatrix} -1 \\ 3 \end{pmatrix}$

   $= \begin{pmatrix} -5 \\ 1 \end{pmatrix}$

   $\overrightarrow{CB} = \overrightarrow{CO} + \overrightarrow{OB} = -\overrightarrow{OC} + \overrightarrow{OB} = \begin{pmatrix} -6 \\ 2 \end{pmatrix} + \begin{pmatrix} -1 \\ 3 \end{pmatrix}$

   $= \begin{pmatrix} -7 \\ 5 \end{pmatrix}$

   $\overrightarrow{AC} = \overrightarrow{AO} + \overrightarrow{OC} = -\overrightarrow{OA} + \overrightarrow{OC} = \begin{pmatrix} -4 \\ -2 \end{pmatrix} + \begin{pmatrix} 6 \\ -2 \end{pmatrix}$

   $= \begin{pmatrix} 2 \\ -4 \end{pmatrix}$

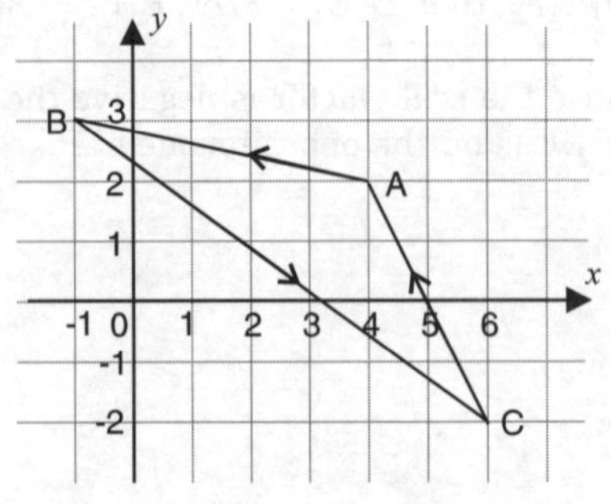

Note: These results could be obtained from the diagram shown. It is always useful to draw a sketch to help you to understand the problem.

EXERCISE 16 (cont.)

3. a) $\overrightarrow{AB} = \overrightarrow{AO} + \overrightarrow{OB} = -\overrightarrow{OA} + \overrightarrow{OB}$
$= -2\underset{\sim}{p} + 2\underset{\sim}{q}$ (or $2\underset{\sim}{q} - 2\underset{\sim}{p}$)

b) $\overrightarrow{ON} = \overrightarrow{OA} + \overrightarrow{AN} = \overrightarrow{OA} + \frac{1}{2}\overrightarrow{AC}$
$= \overrightarrow{OA} + \frac{1}{2}\overrightarrow{OB}$ ($\overrightarrow{AC} = \overrightarrow{OB}$ because ABCD is a parallelogram)
$= 2\underset{\sim}{p} + \underset{\sim}{q}$

c) $\overrightarrow{NM} = \overrightarrow{NO} + \overrightarrow{OB} + \overrightarrow{BM}$ or $\overrightarrow{NC} + \overrightarrow{CM}$
$= -\overrightarrow{ON} + \overrightarrow{OB} + \frac{1}{2}\overrightarrow{BC}$ or $\frac{1}{2}\overrightarrow{AC} + \frac{1}{2}\overrightarrow{CB}$
$= -\overrightarrow{ON} + \overrightarrow{OB} + \frac{1}{2}\overrightarrow{OA}$ or $\frac{1}{2}\overrightarrow{OB} - \frac{1}{2}\overrightarrow{BC}$
$= (-2\underset{\sim}{p} - \underset{\sim}{q}) + 2\underset{\sim}{q} + \underset{\sim}{p}$ or $\frac{1}{2}\overrightarrow{OB} - \frac{1}{2}\overrightarrow{OA}$
$= \underset{\sim}{q} - \underset{\sim}{p}$

4. a) $\overrightarrow{MT} = \frac{1}{2}\overrightarrow{BT} = \frac{1}{2}\overrightarrow{OA} = \frac{1}{2}\underset{\sim}{a}$

b) $\overrightarrow{TN} = \frac{1}{2}\overrightarrow{TA} = \frac{1}{2}\overrightarrow{BO} = -\frac{1}{2}\overrightarrow{OB} = -\frac{1}{2}\underset{\sim}{b}$

c) $\overrightarrow{MN} = \overrightarrow{MT} + \overrightarrow{TN} = \frac{1}{2}\underset{\sim}{a} - \frac{1}{2}\underset{\sim}{b}$

d) Position vector of P $= \overrightarrow{OP} = \overrightarrow{OA} + \overrightarrow{AN} + \overrightarrow{NP}$
$= \overrightarrow{OA} + \frac{1}{2}\overrightarrow{AT} + \frac{1}{2}\overrightarrow{NM}$
$= \overrightarrow{OA} + \frac{1}{2}\overrightarrow{OB} - \frac{1}{2}\overrightarrow{MN}$
$= \underset{\sim}{a} + \frac{1}{2}\underset{\sim}{b} - (\frac{1}{4}\underset{\sim}{a} - \frac{1}{4}\underset{\sim}{b})$
$= \underset{\sim}{a} + \frac{1}{2}\underset{\sim}{b} - \frac{1}{4}\underset{\sim}{a} + \frac{1}{4}\underset{\sim}{b}$
$= \frac{3}{4}\underset{\sim}{a} + \frac{3}{4}\underset{\sim}{b}$

EXERCISE 17

1.

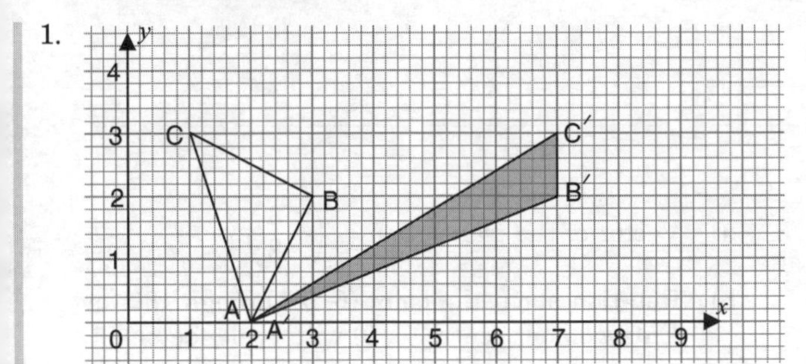

A is on the invariant line so A′ is the same as A.

B is 2 units from the invariant line so it moves 2 × 2 units parallel to the invariant line. Hence, BB′ = 4 units.

C is 3 units from the invariant line so it moves 2 × 3 units parallel to the invariant line. Hence, CC′ = 6 units.

2.

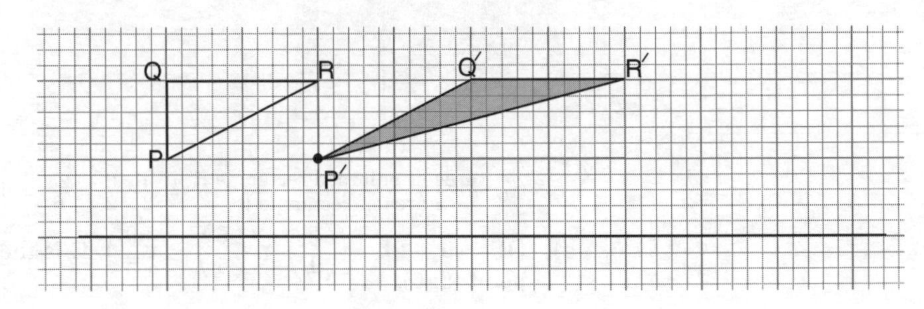

P is 1 cm from the invariant line and it has moved 2 cm to P′. Hence, the shear factor is 2.

Q and R are 2 cm from the invariant line so each of them will move 2 × 2 cm parallel to the invariant line to the positions Q′ and R′ shown.

3. a) Shear with the x-axis as invariant line and shear factor 2.
(C is 2 units from the invariant line and it has moved 4 units to Q.)
b) Stretch with the x-axis invariant and scale factor 2.
(OL = 2 × OC and AK = 2 × AB.)

Check your progress 3

1. a) and b)

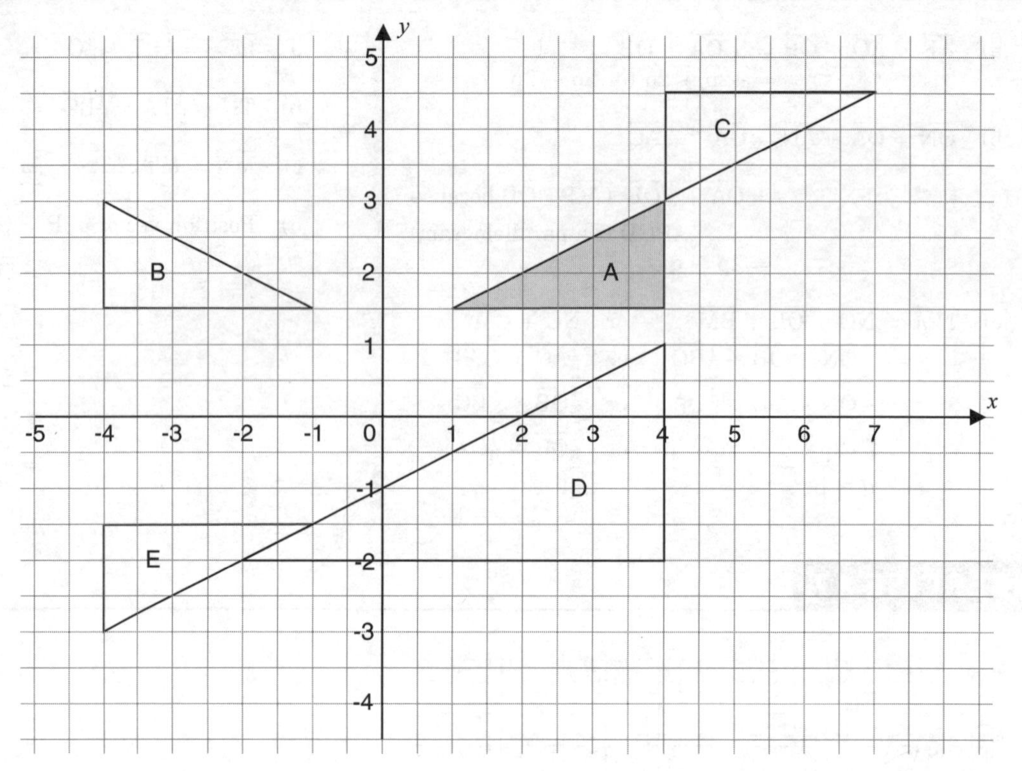

c) A translation with vector $\begin{pmatrix} 8 \\ 6 \end{pmatrix}$.

2. a) (i) $\underset{\sim}{m} + \underset{\sim}{n} = \begin{pmatrix} 3 + (-2) \\ (-4) + 1 \end{pmatrix} = \begin{pmatrix} 1 \\ -3 \end{pmatrix}$

(ii) $3\underset{\sim}{n} = \begin{pmatrix} 3 \times (-2) \\ 3 \times 1 \end{pmatrix} = \begin{pmatrix} -6 \\ 3 \end{pmatrix}$

b)

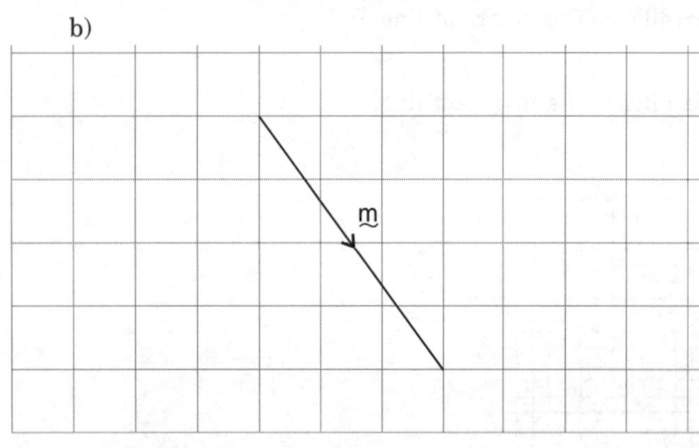

3. a) Reflection in the x-axis.

b) Translation with vector $\begin{pmatrix} -3 \\ 2 \end{pmatrix}$.

c) Enlargement with centre (0, 0) and scale factor 2.

d) Rotation through 90° anticlockwise (or 270° clockwise), centre (0, 0).

4. a) Join A to A′, B to B′, C to C′. These lines meet at the centre of enlargement which is (– 1, 2).

b) $\overrightarrow{A'B'} = -2\overrightarrow{AB}$, $\overrightarrow{B'C'} = -2\overrightarrow{BC}$ and $\overrightarrow{C'A'} = -2\overrightarrow{CA}$ so the scale factor of the enlargement is – 2.

5. a) Vector of the translation =  $\begin{pmatrix} 3 \\ -2 \end{pmatrix}$.

b) Rotation about the point (6, 0) through 180° or Enlargement centre (6, 0) with scale factor –1.

c) (i)

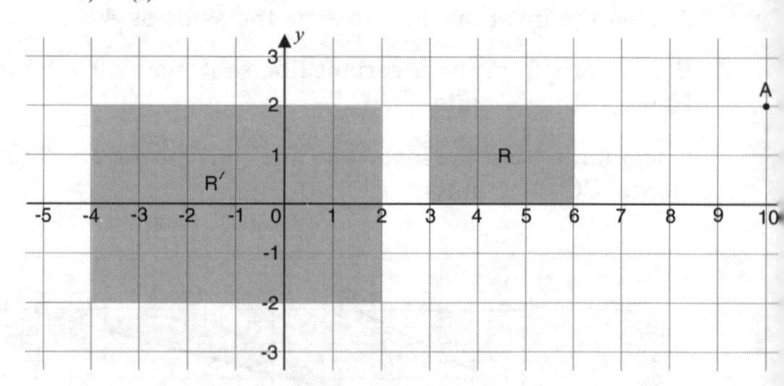

(ii) $\dfrac{\text{area of enlarged rectangle R}'}{\text{area of rectangle R}} = \dfrac{4 \times 6}{3 \times 2} = \dfrac{24}{6} = 4.$

6. a) $\overrightarrow{OC} = \mathbf{a} + 2\mathbf{b} = \begin{pmatrix} 3 \\ 1 \end{pmatrix} + \begin{pmatrix} 6 \\ 4 \end{pmatrix} = \begin{pmatrix} 9 \\ 5 \end{pmatrix}$. C is the point (9, 5).

b) $\overrightarrow{OD} = \overrightarrow{BA} = \overrightarrow{BO} + \overrightarrow{OA} = -\overrightarrow{OB} + \overrightarrow{OA} = -\mathbf{b} + \mathbf{a}$ (or $\mathbf{a} - \mathbf{b}$).

c) $|\mathbf{a}| = \sqrt{3^2 + 1^2} = \sqrt{10} = 3.16$ to 2 decimal places.

7. a) (i) Translation with vector $\begin{pmatrix} 7 \\ 3 \end{pmatrix}$.

(ii) Enlargement with centre (0, 0) and scale factor 3.

(iii) Rotation through 90° anticlockwise (or 270° clockwise) about (0, 0).

(iv) Stretch with the x-axis invariant, scale factor 4.

(v) Shear with the x-axis invariant, shear factor 3.

b) Shapes B, D and F each have area equal to that of shape A.

EXERCISE 18

1. a) 2×2 b) 2×3 c) 3×2
   d) 1×2 e) 1×3 f) 1×1
   g) 2×1 h) 3×1 i) 3×3

2. C = E and H = J.
   Note: Although B has the same elements as A, they are in different positions, so B is not equal to A. The same applies to G and C (or E).

EXERCISE 19

1. $(9 \quad -3)$

2. $\begin{pmatrix} 15 \\ -6 \end{pmatrix}$

3. $\begin{pmatrix} 10 & -6 \\ 0 & 2 \end{pmatrix}$

4. $\begin{pmatrix} -4 & 8 \\ -12 & 0 \end{pmatrix}$

5. $\begin{pmatrix} 2 & 0 \\ 1 & \frac{1}{2} \end{pmatrix}$

6. $\begin{pmatrix} 6a & -6b \\ 9a & 12b \end{pmatrix}$

7. $\begin{pmatrix} -2 & 0 \\ 0 & 2 \end{pmatrix}$

8. $\begin{pmatrix} 8 \\ -2 \end{pmatrix} - \begin{pmatrix} 3 \\ -6 \end{pmatrix} = \begin{pmatrix} 8 - 3 \\ -2 + 6 \end{pmatrix} = \begin{pmatrix} 5 \\ 4 \end{pmatrix}$

EXERCISE 20

1. Remember that (1 by 2 matrix) × (2 by 1 matrix) = (1 by 1 matrix).

   a) $(4 \quad 5)\begin{pmatrix} 3 \\ 2 \end{pmatrix} = (4 \times 3 + 5 \times 2) = (22)$

   b) $(-2 \quad 1)\begin{pmatrix} -3 \\ 0 \end{pmatrix} = ((-2) \times (-3) + 1 \times 0) = (6)$

   c) $(-3 \quad -2)\begin{pmatrix} 3 \\ 4 \end{pmatrix} \; ((-3) \times 3 + (-2) \times 4) = (-17)$

   d) $(3 \quad -2)\begin{pmatrix} -2 \\ 3 \end{pmatrix} = (3 \times (-2) + (-2) \times 3)$
   $= (-12)$

   e) $(a \quad 3)\begin{pmatrix} 1 \\ 3 \end{pmatrix} = (a + 9)$

2. a) $\begin{pmatrix} 5 & 1 \\ 2 & 0 \end{pmatrix}\begin{pmatrix} 2 \\ 3 \end{pmatrix} = \begin{pmatrix} 5 \times 2 + 1 \times 3 \\ 2 \times 2 + 0 \times 3 \end{pmatrix} = \begin{pmatrix} 13 \\ 4 \end{pmatrix}$

   b) $\begin{pmatrix} -1 & -2 \\ -3 & 0 \end{pmatrix}\begin{pmatrix} -2 \\ -3 \end{pmatrix} = \begin{pmatrix} (-1) \times (-2) + (-2) \times (-3) \\ (-3) \times (-2) + 0 + (-3) \end{pmatrix}$
   $= \begin{pmatrix} 8 \\ 6 \end{pmatrix}$

   c) $\begin{pmatrix} 2 & 6 \\ 0 & 3 \end{pmatrix}\begin{pmatrix} 1 & 3 \\ 1 & 4 \end{pmatrix}$
   $= \begin{pmatrix} 2 \times 1 + 6 \times 1 & 2 \times 3 + 6 \times 4 \\ 0 \times 1 + 3 \times 1 & 0 \times 3 + 3 \times 4 \end{pmatrix} = \begin{pmatrix} 8 & 30 \\ 3 & 12 \end{pmatrix}$

   d) $\begin{pmatrix} 1 & -2 \\ 3 & 5 \end{pmatrix}\begin{pmatrix} 3 & -1 \\ -1 & 1 \end{pmatrix}$
   $= \begin{pmatrix} 1 \times 3 + (-2) \times (-1) & 1 \times (-1) + (-2) \times 1 \\ 3 \times 3 + 5 \times (-1) & 3 \times (-1) + 5 \times 1 \end{pmatrix}$
   $= \begin{pmatrix} 5 & -3 \\ 4 & 2 \end{pmatrix}$

3. a) $(x + 6) = (10)$ so $x = 4$
   b) $(15 + 2x) = (17)$ so $x = 1$
   c) $(2x - 15) = (-1)$ so $x = 7$

4. a) $AB = \begin{pmatrix} 0 & 5 \\ 2 & -2 \end{pmatrix}\begin{pmatrix} 1 & 3 \\ -1 & 2 \end{pmatrix}$
   $= \begin{pmatrix} 0 \times 1 + 5(-1) & 0 \times 3 + 5 \times 2 \\ 2 \times 1 + (-2)(-1) & 2 \times 3 + (-2)(2) \end{pmatrix}$
   $= \begin{pmatrix} -5 & 10 \\ 4 & 2 \end{pmatrix}$

   b) $BA = \begin{pmatrix} 1 & 3 \\ -1 & 2 \end{pmatrix}\begin{pmatrix} 0 & 5 \\ 2 & -2 \end{pmatrix}$
   $= \begin{pmatrix} 1 \times 0 + 3 \times 2 & 1 \times 5 + (3)(-2) \\ (-1)(0) + 2 \times 2 & (-1)(5) + (2)(-2) \end{pmatrix}$
   $= \begin{pmatrix} 6 & -1 \\ 4 & -9 \end{pmatrix}$

   Note: BA is not equal to AB.

   c) $AI = \begin{pmatrix} 0 & 5 \\ 2 & -2 \end{pmatrix}\begin{pmatrix} 1 & 0 \\ 0 & 1 \end{pmatrix} = \begin{pmatrix} 0 & 5 \\ 2 & -2 \end{pmatrix}$

   d) $IA = \begin{pmatrix} 1 & 0 \\ 0 & 1 \end{pmatrix}\begin{pmatrix} 0 & 5 \\ 2 & -2 \end{pmatrix} = \begin{pmatrix} 0 & 5 \\ 2 & -2 \end{pmatrix}$

   Note: IA is equal to AI (and each of them is equal to A).

   e) $B^2 = \begin{pmatrix} 1 & 3 \\ -1 & 2 \end{pmatrix}\begin{pmatrix} 1 & 3 \\ -1 & 2 \end{pmatrix}$
   $= \begin{pmatrix} 1 \times 1 + (3)(-1) & 1 \times 3 + 3 \times 2 \\ (-1)(1) + (2)(-1) & (-1)(3) + 2 \times 2 \end{pmatrix}$
   $= \begin{pmatrix} -2 & 9 \\ -3 & 1 \end{pmatrix}$

   $B^3 = B^2 \times B = \begin{pmatrix} -2 & 9 \\ -3 & 1 \end{pmatrix}\begin{pmatrix} 1 & 3 \\ -1 & 2 \end{pmatrix}$
   $= \begin{pmatrix} (-2)(1) + (9)(-1) & (-2)(3) + 9 \times 2 \\ (-3)(1) + (1)(-1) & (-3)(3) + 1 \times 2 \end{pmatrix}$
   $= \begin{pmatrix} -11 & 12 \\ -4 & -7 \end{pmatrix}$

   Check: Show that $B \times B^2$ gives the same result as $B^2 \times B$.

EXERCISE 21

1. a) $(4 \times 5) - (1 \times 3) = 20 - 3 = 17$
   b) $(6 \times 4) - (8 \times 3) = 24 - 24 = 0$
   c) $(2 \times 9) - ((-1) \times 3) = 18 + 3 = 21$
   d) $((-1) \times 1) - (0 \times 0) = -1 - 0 = -1$
   e) $((-3) \times 1) - ((-2) \times (-1)) = -3 - 2 = -5$

2. $x - 6 = 4$ so $x = 10$

3. $x + 8 = 5$ so $x = -3$

4. $2y - 6 = 2$ so $y = 4$

5. $-6 + 2p = 12$ so $p = 9$

EXERCISE 22

1. $\begin{pmatrix} 1 & -2 \\ -1 & 3 \end{pmatrix}$ 2. $\begin{pmatrix} 2 & -1 \\ -1 & 1 \end{pmatrix}$ 3. $\begin{pmatrix} 7 & -4 \\ -5 & 3 \end{pmatrix}$ 4. $\begin{pmatrix} 3 & 7 \\ 2 & 5 \end{pmatrix}$ 5. $\begin{pmatrix} 3 & 2 \\ 4 & 3 \end{pmatrix}$

EXERCISE 23

1. a) Determinant $= 4 \times 2 - 6 \times 1 = 8 - 6 = 2$

   Inverse matrix $= \frac{1}{2}\begin{pmatrix} 2 & -6 \\ -1 & 4 \end{pmatrix} = \begin{pmatrix} 1 & -3 \\ -\frac{1}{2} & 2 \end{pmatrix}$

   b) Determinant $= 2 \times 8 - 3 \times 5 = 1$

   Inverse matrix $= \begin{pmatrix} 8 & -3 \\ -5 & 2 \end{pmatrix}$

   c) Determinant $= 3 \times 4 - 2 \times 6 = 0$
   The matrix does not have an inverse.

   d) Determinant $= (3) \times (-1) - (-2) \times (2) = -3 + 4 = 1$

   Inverse matrix $= \begin{pmatrix} -1 & 2 \\ -2 & 3 \end{pmatrix}$

   e) Determinant $= 3 \times 6 - 4 \times 5 = 18 - 20 = -2$

   Inverse matrix $= -\frac{1}{2}\begin{pmatrix} 6 & -4 \\ -5 & 3 \end{pmatrix} = \begin{pmatrix} -3 & 2 \\ 2.5 & -1.5 \end{pmatrix}$

2. Determinant of $I = 1 \times 1 - 0 \times 0 = 1$

   $I^{-1} = \frac{1}{1}\begin{pmatrix} 1 & -0 \\ -0 & 0 \end{pmatrix} = \begin{pmatrix} 1 & 0 \\ 0 & 1 \end{pmatrix} = I$

3. a) Determinant $= (1)(-1) - (0)(0) = -1 - 0 = -1$

   Inverse matrix $= -\frac{1}{1}\begin{pmatrix} -1 & -0 \\ -0 & 1 \end{pmatrix} = \begin{pmatrix} 1 & 0 \\ 0 & -1 \end{pmatrix}$

   = the given matrix

   b) Determinant $= (-1)(-1) - (0)(0) = +1 - 0 = 1$

   Inverse matrix $= \frac{1}{1}\begin{pmatrix} -1 & -0 \\ -0 & -1 \end{pmatrix} = \begin{pmatrix} -1 & 0 \\ 0 & -1 \end{pmatrix}$

   = the given matrix

   c) Determinant $= (0)(0) - (1)(1) = 0 - 1 = -1$

   Inverse matrix $= -\frac{1}{1}\begin{pmatrix} 0 & -1 \\ -1 & 0 \end{pmatrix} = \begin{pmatrix} 0 & 1 \\ 1 & 0 \end{pmatrix}$

   = the given matrix

   d) Determinant $= (0)(0) - (-1)(-1) = 0 - 1 = -1$

   Inverse matrix $= -\frac{1}{1}\begin{pmatrix} 0 & 1 \\ 1 & 0 \end{pmatrix} = \begin{pmatrix} 0 & -1 \\ -1 & 0 \end{pmatrix}$

   = the given matrix

4. a) $|A| = 3 \times 5 - 7 \times 2 = 15 - 14 = 1$

   $A^{-1} = \begin{pmatrix} 5 & -7 \\ -2 & 3 \end{pmatrix}$

   b) The equations are equivalent to

   $\begin{pmatrix} 3 & 7 \\ 2 & 5 \end{pmatrix}\begin{pmatrix} x \\ y \end{pmatrix} = \begin{pmatrix} 5 \\ 3 \end{pmatrix}$

   Multiplying each side by A^{-1} on the left:

   $\begin{pmatrix} 5 & -7 \\ -2 & 3 \end{pmatrix}\begin{pmatrix} 3 & 7 \\ 2 & 5 \end{pmatrix}\begin{pmatrix} x \\ y \end{pmatrix} = \begin{pmatrix} 5 & -7 \\ -2 & 3 \end{pmatrix}\begin{pmatrix} 5 \\ 3 \end{pmatrix}$

   Hence $\begin{pmatrix} 1 & 0 \\ 0 & 1 \end{pmatrix}\begin{pmatrix} x \\ y \end{pmatrix} = \begin{pmatrix} (5)(5) + (-7)(3) \\ (-2)(5) + (3)(3) \end{pmatrix}$

   $\begin{pmatrix} x \\ y \end{pmatrix} = \begin{pmatrix} 4 \\ -1 \end{pmatrix}$

   The solution is $x = 4$, $y = -1$.

EXERCISE 24

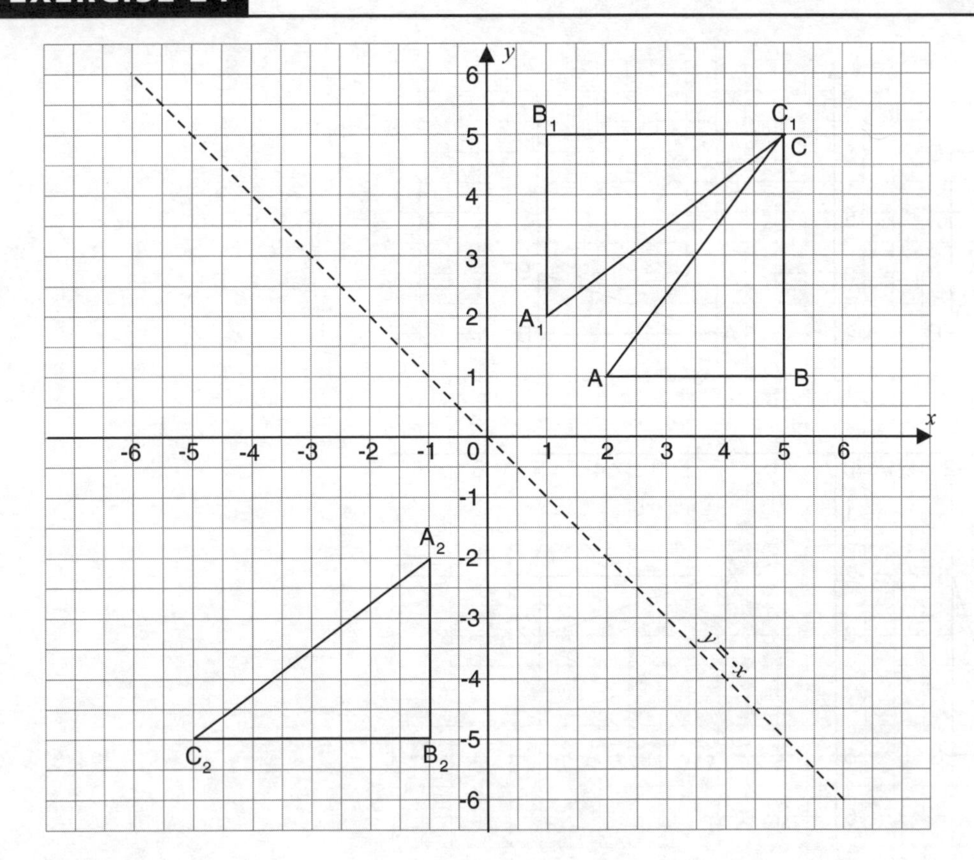

a) See graph.

b) (i) See graph. $\begin{pmatrix} 0 & 1 \\ 1 & 0 \end{pmatrix} \overset{\begin{matrix} A & B & C \end{matrix}}{\begin{pmatrix} 2 & 5 & 5 \\ 1 & 1 & 5 \end{pmatrix}} = \overset{\begin{matrix} A_1 & B_1 & C_1 \end{matrix}}{\begin{pmatrix} 1 & 1 & 5 \\ 2 & 5 & 5 \end{pmatrix}}$

(ii) Reflection in the line $y = x$.

c) (i) See graph.

(ii) Reflection in the line $y = -x$ is represented by the matrix $\begin{pmatrix} 0 & -1 \\ -1 & 0 \end{pmatrix}$

(The image of the point with position vector $\begin{pmatrix} 1 \\ 0 \end{pmatrix}$ is $\begin{pmatrix} 0 \\ -1 \end{pmatrix}$ and the image of the point with position vector $\begin{pmatrix} 0 \\ 1 \end{pmatrix}$ is $\begin{pmatrix} -1 \\ 0 \end{pmatrix}$.)

d) (i) Rotation about the origin through 180° or

Enlargement with scale factor -1 and centre the origin.

(ii) The transformation is represented by the matrix $\begin{pmatrix} -1 & 0 \\ 0 & -1 \end{pmatrix}$.

(The image of $\begin{pmatrix} 1 \\ 0 \end{pmatrix}$ is $\begin{pmatrix} -1 \\ 0 \end{pmatrix}$ and the image of $\begin{pmatrix} 0 \\ 1 \end{pmatrix}$ is $\begin{pmatrix} 0 \\ -1 \end{pmatrix}$.)

EXERCISE 24 (cont.)

2.

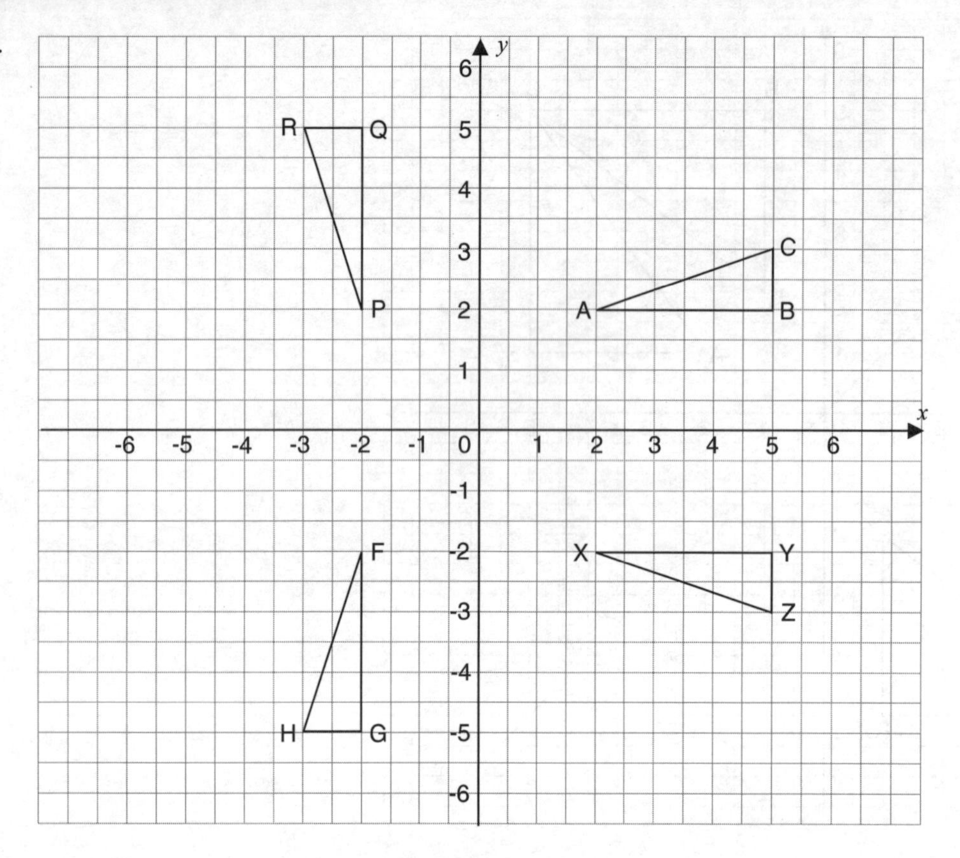

a) See graph.

b)

$$\begin{pmatrix} 0 & -1 \\ 1 & 0 \end{pmatrix} \overset{\quad A \quad B \quad C}{\begin{pmatrix} 2 & 5 & 5 \\ 2 & 2 & 3 \end{pmatrix}} = \overset{\quad P \quad Q \quad R}{\begin{pmatrix} -2 & -2 & -3 \\ 2 & 5 & 5 \end{pmatrix}}$$

Triangle PQR is shown on the graph.

c)

$$\begin{pmatrix} 1 & 0 \\ 0 & -1 \end{pmatrix} \overset{\quad A \quad B \quad C}{\begin{pmatrix} 2 & 5 & 5 \\ 2 & 2 & 3 \end{pmatrix}} = \overset{\quad X \quad Y \quad Z}{\begin{pmatrix} 2 & 5 & 5 \\ -2 & -2 & -3 \end{pmatrix}}$$

Triangle XYZ is shown on the graph.

d) (i) Reflection in the line $y = x$.

(ii) The image of $\begin{pmatrix} 1 \\ 0 \end{pmatrix}$ is $\begin{pmatrix} 0 \\ 1 \end{pmatrix}$ and the image of $\begin{pmatrix} 0 \\ 1 \end{pmatrix}$ is $\begin{pmatrix} 1 \\ 0 \end{pmatrix}$ and so the matrix which represents the transformation is $\begin{pmatrix} 0 & 1 \\ 1 & 0 \end{pmatrix}$.

e) (i) $NM = \begin{pmatrix} 1 & 0 \\ 0 & -1 \end{pmatrix}\begin{pmatrix} 0 & -1 \\ 1 & 0 \end{pmatrix} = \begin{pmatrix} 0 & -1 \\ -1 & 0 \end{pmatrix}$

(ii)

$$\begin{pmatrix} 0 & -1 \\ -1 & 0 \end{pmatrix} \overset{\quad A \quad B \quad C}{\begin{pmatrix} 2 & 5 & 5 \\ 2 & 2 & 3 \end{pmatrix}} = \overset{\quad F \quad G \quad H}{\begin{pmatrix} -2 & -2 & -3 \\ -2 & -5 & -5 \end{pmatrix}}$$

Triangle FGH is shown on the graph.

(iii) Transformation T followed by transformation U.

Check your progress 4

1. a) (ii) AC is possible (1 by 2 matrix) × (2 by 2 matrix) = (1 by 2 matrix).

   b) $AB = (1 \quad 2)\begin{pmatrix} -3 \\ 4 \end{pmatrix} = ((1)(-3) + (2)(4)) = (5)$

   c) $|C| = (-2)(6) - (5)(-3) = -12 + 15 = 3$

   $$C^{-1} = \frac{1}{3}\begin{pmatrix} 6 & -5 \\ 3 & -2 \end{pmatrix} = \begin{pmatrix} 2 & -\frac{5}{3} \\ 1 & -\frac{2}{3} \end{pmatrix}$$

2. a) $\begin{pmatrix} 3 & 2 \\ -1 & 6 \end{pmatrix}\begin{pmatrix} -3 \\ 2 \end{pmatrix} = \begin{pmatrix} (3)(-3) + (2)(2) \\ (-1)(-3) + (6)(2) \end{pmatrix} = \begin{pmatrix} -5 \\ 15 \end{pmatrix}$

   Hence, $x = -5$ and $y = 15$.

   b) Determinant of $\begin{pmatrix} 2 & -1 \\ 4 & 3 \end{pmatrix} = (2)(3) - (-1)(4) = 6 + 4 = 10$

   Inverse of $\begin{pmatrix} 2 & -1 \\ 4 & 3 \end{pmatrix} = \frac{1}{10}\begin{pmatrix} 3 & 1 \\ -4 & 2 \end{pmatrix} = \begin{pmatrix} 0.3 & 0.1 \\ -0.4 & 0.2 \end{pmatrix}$

   c) $\begin{pmatrix} (3t)(1) + (u)(2) \\ (-t)(1) + (3u)(2) \end{pmatrix} = \begin{pmatrix} 10 \\ -10 \end{pmatrix}$ which is $\begin{pmatrix} 3t + 2u \\ -t + 6u \end{pmatrix} = \begin{pmatrix} 10 \\ -10 \end{pmatrix}$

   Hence, $3t + 2u = 10$ and $-t + 6u = -10$

   To solve these simultaneous equations, multiply the second one by 3.

   $3t + 2u = 10$ and $-3t + 18u = -30$.

   Adding these equations gives $20u = -20$ and so $u = -1$.
   Substituting this in either of the equations gives $t = 4$.
   The solution of the equations is $t = 4$, $u = -1$.

3. a)

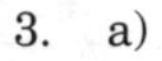

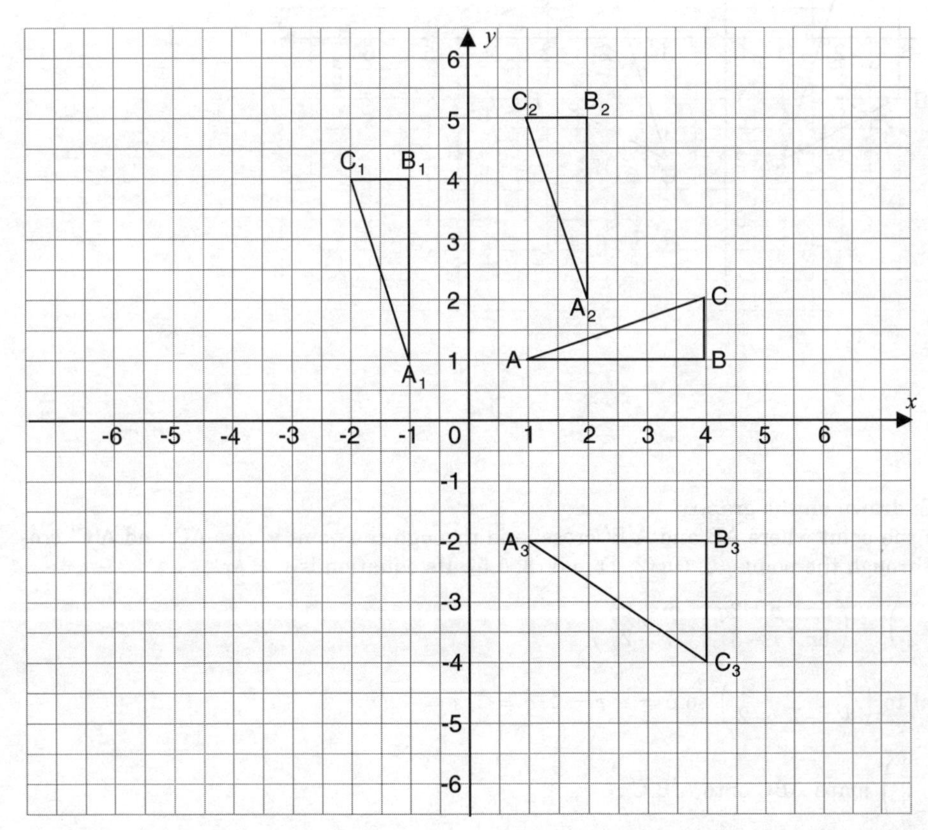

Check your progress 4 (cont.)

b) (i) See graph.
(ii) The position vectors of A_1, B_1, C_1 are $\begin{pmatrix}-1\\1\end{pmatrix}$, $\begin{pmatrix}-1\\4\end{pmatrix}$, $\begin{pmatrix}-2\\4\end{pmatrix}$.

Under the translation, the image of $\begin{pmatrix}x\\y\end{pmatrix}$ is $\begin{pmatrix}x+3\\y+1\end{pmatrix}$.

The position vectors of A_2, B_2, C_2 are $\begin{pmatrix}2\\2\end{pmatrix}$, $\begin{pmatrix}2\\5\end{pmatrix}$, $\begin{pmatrix}1\\5\end{pmatrix}$.

(iii) Rotation through 90° anticlockwise about the point (1, 2).

c) (i)

$$\begin{pmatrix}1 & 0\\0 & -2\end{pmatrix}\overset{\quad A\ \ B\ \ C}{\begin{pmatrix}1 & 4 & 4\\1 & 1 & 2\end{pmatrix}} = \overset{A_3\ \ B_3\ \ C_3}{\begin{pmatrix}1 & 4 & 4\\-2 & -2 & -4\end{pmatrix}}$$

Triangle $A_3B_3C_3$ is drawn on the graph.
(ii) Stretch with the x-axis invariant and scale factor −2.

4.

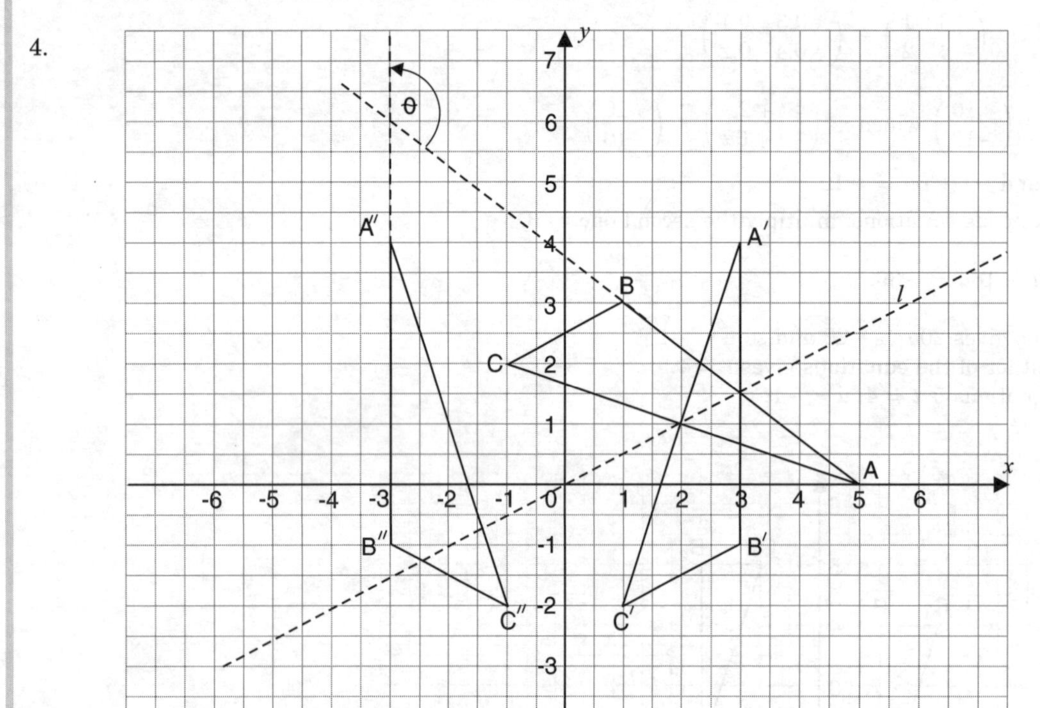

a) See the graph.
b) (i) The mirror line l is drawn on the graph.
(It passes through the point where AB and A′B′ cross, and through the point where AC and A′C′ cross.)
(ii) The line l passes through the points (0, 0), (2, 1) and (3, 1.5). Its equation is $y = \frac{1}{2}x$.

(iii) $\begin{pmatrix}p & q\\r & s\end{pmatrix}\begin{pmatrix}5 & 1 & -1\\0 & 3 & 2\end{pmatrix} = \begin{pmatrix}5p & p+3q & -p+2q\\5r & r+3s & -r+2s\end{pmatrix}$

This must be equal to $\begin{pmatrix}3 & 3 & 1\\4 & -1 & -2\end{pmatrix}$ so $p = \frac{3}{5}$, $r = \frac{4}{5}$, $q = \frac{4}{5}$, $s = -\frac{3}{5}$.

(iv) The matrix $\begin{pmatrix}\frac{3}{5} & \frac{4}{5}\\\frac{4}{5} & -\frac{3}{5}\end{pmatrix}$ maps ABC onto A′B′C′.

It represents reflection in the line $y = \frac{1}{2}x$.

c) See the graph.
d) The angle of rotation is the angle between $\overrightarrow{AB}$ and $\overrightarrow{A''B''}$. (Angle θ in the diagram). By measurement, it is 127°. (Answers between 126° and 128° acceptable.)

Index